Name _____ Class ____

Skills Worksheet
Directed Reading

Section: How Populations Grow

Complete each statement by writing the correct term or phrase in the space provided.

1. A(n) _____ consists of all the individuals of a species that live together in one place at one time.

2. The statistical study of all populations is called _____ .

3. The way individuals of a population are arranged in space is called _____ .

4. Demographers are interested in predicting how the _____ of a population will change.

Read each question, and write your answer in the space provided.

5. Why is size an important feature of a population?

_____

_____

_____

_____

_____

_____

6. Why is density an important feature of a population?

_____

_____

_____

_____

_____

_____

_____

Copyright © by Holt, Rinehart and Winston. All rights reserved.

Holt Biology · Populations

Name _____ Class _____ Date _____

Directed Reading continued

7. What is a population model?

_____
_____
_____

8. How is the growth rate of a population calculated?

_____
_____
_____
_____

9. What are density-dependent factors? Give an example of a density-dependent factor.

_____
_____
_____
_____

10. What is the logistic model of population growth?

_____
_____
_____
_____

Name _____ Class _____ Date _____

Directed Reading continued

Complete each statement by writing the correct term or phrase in the space provided.

11. In a(n) _____ _____ _____ , the rate of population growth is constant.

12. The population size that an environment can sustain is called the

_____ _____ .

In the space provided, write the letter of the description that best matches the term or phrase.

_____ **13.** density-independent factors

_____ **14.** r-strategists

_____ **15.** K-strategists

_____ **16.** density-dependent factors

a. grow exponentially when environmental conditions allow them to reproduce

b. environmental conditions

c. their population density is usually near the carrying capacity of their environment

d. amount of limited resources

In the space provided, write the letter of the term or phrase that best completes each statement or best answers each question.

_____ **17.** Which of the following are mosquito populations?
  a. r-strategists
  b. K-strategists
  c. N-strategists
  d. pq-strategists

_____ **18.** Many r-strategists are found in a
  a. stable environment.
  b. rapidly changing environment.
  c. humid environment.
  d. slowly changing environment.

_____ **19.** In general, r-strategists have all of the following characteristics EXCEPT
  a. a short life span.
  b. reproduction early in life.
  c. parental care of offspring.
  d. production of many offspring.

_____ **20.** Populations that grow slowly are
  a. K-strategists.
  b. endangered.
  c. genetically similar.
  d. decreasing in allele frequency.

Name _____ Class _____ Date _____

Skills Worksheet

Directed Reading

Section: How Populations Evolve

In the space provided, write the letter of the description that best matches the term or phrase.

_____ 1. natural selection

_____ 2. gene flow

_____ 3. genetic drift

_____ 4. nonrandom mating

_____ 5. mutation

a. the source of variation that makes evolution possible

b. a change in allele frequency because of random occurrences

c. individuals show preferences in the choice of breeding partners

d. one of the most powerful agents of genetic change

e. the movement of alleles into or out of a population

In the space provided, write the letter of the term or phrase that best completes each statement or best answers each question.

_____ 6. The Hardy-Weinberg principle states that
   a. an equation can be used to predict phenotype frequencies in a population.
   b. the frequencies of alleles in a population do not change unless evolutionary forces act on the population.
   c. dominant alleles automatically replace recessive alleles.
   d. populations change in the absence of evolutionary forces.

_____ 7. Small populations are more likely to undergo which of the following?
   a. natural selection
   b. mutation
   c. genetic drift
   d. gene flow

_____ 8. When a female bird prefers to mate with males with long tail feathers, this behavior is called
   a. genetic drift.
   b. natural selection.
   c. nonrandom mating.
   d. gene flow.

_____ 9. Evolutionary forces acting in a population can cause allele frequencies to
   a. change.
   b. stabilize.
   c. disappear.
   d. be expressed.

_____ 10. An allele can increase or decrease in frequency, depending on how that change affects a species'
   a. survival and reproduction.
   b. habitat.
   c. carrying capacity.
   d. dispersion.

Name _____ Class _____ Date _____

Directed Reading *continued*

Read each question, and write your answer in the space provided.

11. Explain why natural selection does not act directly on genes.

_____

_____

_____

_____

12. Given that people with hemophilia rarely reproduce, why has natural selection not eliminated hemophilia from the population?

_____

_____

_____

_____

Complete each statement by writing the correct term or phrase in the space provided.

13. A trait that is influenced by several genes is called a(n)

_____ trait.

14. A hill-shaped curve centered around an average value is called a(n)

_____ _____ .

Read each question, and write your answer in the space provided.

15. What is directional selection?

_____

_____

_____

_____

16. What is stabilizing selection?

_____

_____

Name _____ Class _____ Date _____

Skills Worksheet

Active Reading

Section: How Populations Grow
Read the passage below. Then answer the questions that follow.

Every population has features that help determine its future. One of the most important features of any population is its size. The number of individuals in a population, or **population size,** can affect the population's ability to survive. Studies have shown that very small populations are among those most likely to become extinct.

A second important feature of a population is its density. **Population density** is the number of individuals that live in a given area. If the individuals of a population are few and spaced widely apart, they may seldom encounter one another, making reproduction rare.

A third feature of a population is the way the individuals of the population are arranged in space. This feature is called **dispersion.** Three main patterns of dispersion are possible within a population. If the individuals are randomly spaced, the location of each individual is self-determined. If individuals are evenly spaced, they are located at regular intervals. In a clumped distribution, individuals are bunched together in clusters. Each of these patterns reflects the interactions between the population and its environment.

SKILL: READING EFFECTIVELY

Read each question, and write your answer in the space provided.

1. What are three key features of a population?

_____

_____

2. What do studies indicate about very small populations?

_____

_____

3. What is population density?

_____

_____

Name _____ Class _____ Date _____

Active Reading *continued*

4. Describe a situation in which population density has a negative impact on the production of offspring.

_____

_____

_____

SKILL: INTERPRETING GRAPHICS

The figures below show three possible patterns of dispersion in a population. Describe each pattern in the spaces provided.

Pattern a   Pattern b   Pattern c

5. Pattern a: _____

_____

6. Pattern b: _____

_____

7. Pattern c: _____

_____

In the space provided, write the letter of the term or phrase that best completes the statement.

_____ 8. The patterns of dispersion illustrated in the figures above are similar in that they all reflect interactions between
   a. the population and its environment.
   b. producers and consumers.
   c. a population and its members.
   d. Both (a) and (b)

Name _____ Class _____ Date _____

Skills Worksheet
Active Reading

Section: How Populations Evolve

Read the passage below. Then answer the questions that follow.

In 1908, the English mathematician G. H. Hardy and the German physician Wilhelm Weinberg independently demonstrated that dominant alleles do not automatically replace recessive alleles. Using algebra and a simple application of the theories of probability, they showed that the frequency of alleles in a population and the ratio of heterozygous individuals to homozygous individuals does not change from generation to generation unless the population is acted on by other processes that favor particular alleles. Their discovery, called the **Hardy-Weinberg principle,** states that the frequencies of alleles in a population do not change unless evolutionary forces act on the population.

The Hardy-Weinberg principle holds true for any population as long as the population is large enough that its members are not likely to mate with relatives and as long as evolutionary forces are not acting. There are five principal evolutionary forces: mutation, gene flow, nonrandom mating, genetic drift, and natural selection. These evolutionary forces can cause the ratios of genotypes in a population to differ significantly from those predicted by the Hardy-Weinberg principle.

SKILL: READING EFFECTIVELY

Read each question, and write your answer in the space provided.

1. What did Hardy and Weinberg independently demonstrate in 1908?

_____

_____

2. According to the Hardy-Weinberg principle, what causes a change in the frequencies of alleles in a population?

_____

_____

Copyright © by Holt, Rinehart and Winston. All rights reserved.

Holt Biology — Populations

Name _____ Class _____ Date _____

Active Reading *ontinued*

3. What are the five principal evolutionary forces?

   _____

   _____

4. What effect do these evolutionary forces have on the ratio of heterozygous and homozygous individuals in a population?

   _____

   _____

5. How can the population size cause a change in the frequencies of alleles in the population?

   _____

   _____

   _____

In the space provided, write the letter of the term or phrase that best completes the statement.

_____ 6. In forming their theories, Hardy and Weinberg used
   a. simple algebra.
   b. theories of probability.
   c. analytic geometry.
   d. Both (a) and (b)

Name _____ Class _____ Date _____

Skills Worksheet

Vocabulary Review

Complete each statement by writing the correct term or phrase from the list below in the space provided.

carrying capacity   genetic drift   population
density-dependent factors   Hardy-Weinberg principle   population density
density-independent factors   *K*-strategists   population model
directional selection   logistic model   population size
dispersion   nonrandom mating   *r*-strategists
exponential growth curve   normal distribution   stabilizing selection
gene flow   polygenic trait

1. A(n) _____ consists of all the individuals of a species that live together in one place at one time.

2. One of the most important features of any population is its _____ _____, the number of individuals in a population.

3. A second important feature of a population is _____ _____, the number of individuals that live in a given area.

4. A third feature of populations is _____, which refers to the way the individuals of the population are arranged in space.

5. When demographers try to predict how a population will grow, they use a(n) _____ _____, a hypothetical population that exhibits the key characteristics of a real population.

6. When the rate of population growth stays the same and population size is plotted against time on a graph, the population growth curve resembles a J-shaped curve called a(n) _____ _____ _____.

7. The population that an environment can sustain is called the _____ _____.

Copyright © by Holt, Rinehart and Winston. All rights reserved.

Holt Biology — Populations

Name _____ Class _____ Date _____

Vocabulary Review *continued*

8. As populations grow, limited resources get used up. These resources are called _____-_____ _____ because the rate at which they become depleted depends on the density of the population that uses them.

9. The _____ _____ is a population growth model in which exponential growth is limited by a density-dependent factor.

10. Many species of plants and insects reproduce rapidly. Their growth is usually limited by environmental conditions, also known as _____-_____ _____ .

11. Many of these species grow exponentially when environmental conditions permit their reproduction. Such species are called _____ .

12. Slow-growing populations, such as whales and redwood trees, are called _____ because their population density is usually near the carrying capacity (K) of their environment.

13. According to the _____-_____ _____ , the frequencies of alleles in a population do not change unless evolutionary forces act on the population.

14. The evolutionary forces include the mutation of genes and _____ _____ , which is the movement of alleles into or out of a population.

15. Sometimes individuals prefer to mate with others that live nearby or are of their own phenotype, a situation called _____ _____ .

16. In small populations, the frequency of an allele can be greatly changed by a chance event, such as a fire or landslide. This change in allele frequency is called _____ _____ .

Name _____ Class _____ Date _____

Vocabulary Review continued

17. A trait that is influenced by several genes is called a(n) _____ _____ .

18. If you were to plot the height of everyone in your class on a graph, the values would probably form a hill-shaped curve called a(n) _____ _____ .

19. When selection causes the frequency of a particular trait to move in one direction, this form of selection is called _____ _____ .

20. When selection eliminates extremes at both ends of a range of phenotypes, the frequencies of the intermediate phenotypes increase. This form of selection is called _____ _____ .

Name _____ Class _____ Date _____

Skills Worksheet

Science Skills

Interpreting Graphs

Use the graph below, which shows the growth of a population over time, to answer questions 1 and 2.

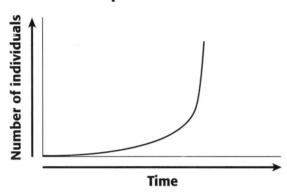

Population Growth

Read each question, and write your answer in the space provided.

1. What type of growth pattern is shown in the graph above? Describe the rate of population growth in this growth pattern.

   _____
   _____
   _____
   _____

2. What types of organisms (r-strategists or K-strategists) are represented by this growth pattern? Describe the features that lead to this growth pattern.

   _____
   _____
   _____
   _____
   _____
   _____
   _____

Copyright © by Holt, Rinehart and Winston. All rights reserved.

Holt Biology   Populations

Name _____ Class _____ Date _____

Science Skills *continued*

Use the graph below, which shows the frequency of a certain allele in populations of different sizes, to answer questions 3–6.

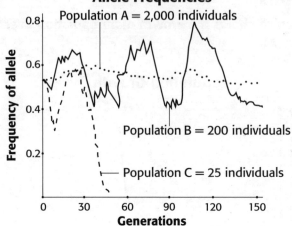

3. In which of these populations does the Hardy-Weinberg principle apply? Explain.

_____
_____
_____

4. What happens to the frequency of the allele in the smallest population shown in the graph above? Explain.

_____
_____
_____

5. What conditions must be present in order for the Hardy-Weinberg principle to apply?

_____
_____
_____

6. What factors might cause genotype frequencies to deviate from those predicted by the Hardy-Weinberg equation?

_____
_____

Name _____ Class _____ Date _____

Skills Worksheet
Concept Mapping

Using the terms and phrases provided below, complete the concept map showing the characteristics of populations.

- carrying capacity
- density-independent factors
- exponential growth curve
- growth rate
- *K*-strategists
- population density
- population models
- population size
- *r*-strategists

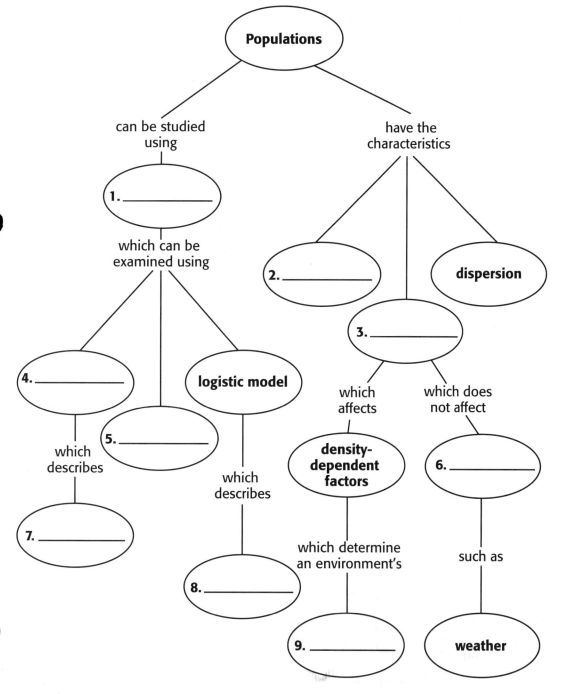

Name _____ Class _____ Date _____

Skills Worksheet
Critical Thinking

Work-Alikes

In the space provided, write the letter of the term or phrase that best describes how each numbered item functions.

_____ 1. model

_____ 2. *K*-strategist

_____ 3. Hardy-Weinberg principle

_____ 4. polygenic trait

_____ 5. carrying capacity

_____ 6. normal distribution

a. math formula

b. survival specialist

c. "king-of-the-hill" game

d. simulator

e. weight limit for a bridge

f. soup with many different ingredients

Cause and Effect

In the space provided, write the letter of the term or phrase that best matches each cause or effect given below.

Cause	Effect
7. _____	individuals homozygous for harmful recessive traits
8. individuals are in randomly spaced populations	_____
9. _____	accumulation of wastes
10. _____	exponential growth is limited
11. density-independent factors	_____
12. _____	inbreeding
13. fire or landslide effect on population	_____
14. allele common enough to produce homozygous offspring	_____

a. nonrandom mating

b. genetically uniform population

c. mosquitoes die in fall or winter

d. population approaching carrying capacity

e. density-dependent factors

f. genetic drift

g. individuals are self-determined as to their location

h. natural selection can act

Name _____ Class _____ Date _____

Critical Thinking continued

Linkages

In the spaces provided, write the letters of the two terms or phrases that are linked together by the term or phrase in the middle. The choices can be placed in any order.

15. _____ population _____
16. _____ population density _____
17. _____ exponential curve _____
18. _____ K-strategists _____
19. _____ directional selection _____
20. _____ stabilizing selection _____

a. population size
b. one extreme from phenotypic range is eliminated
c. logistic model
d. stable environments
e. distribution becomes narrower
f. slow-growing
g. extremes from both ends of phenotypic range are eliminated
h. demography
i. alleles for this become less common
j. pairs of individuals have multiple offspring over lifetime
k. stage I (birth rate – death rate)
l. dispersion

Analogies

An analogy is a relationship between two pairs of terms or phrases written as a : b :: c : d. The symbol : is read as "is to," and the symbol :: is read as "as." In the space provided, write the letter of the pair of terms or phrases that best completes the analogy shown.

_____ 21. clumped dispersions : cluster ::
   a. geese flying in formation : circle
   b. geese flying in formation : V-shape
   c. soldiers in line : cluster
   d. seats in a row : triangle

_____ 22. r : rate of growth ::
   a. p : recessive allele
   b. K : carrying capacity
   c. q : dominant allele frequency
   d. A : recessive allele

_____ 23. weed growth : tree growth ::
   a. carrying capacity : birth rate
   b. logistic growth curve : exponential growth curve
   c. K-strategist population : exponential growth
   d. r-strategist population : K-strategist population

Name _____ Class _____ Date _____

Assessment
Test Prep Pretest

In the space provided, write the letter of the term or phrase that best completes each statement or best answers each question.

_____ 1. The three main patterns of dispersion in a population are
   a. nonrandomly spaced, evenly spaced, and clumped distribution.
   b. nonrandomly spaced, evenly spaced, and unevenly spaced.
   c. randomly spaced, evenly spaced, and clumped distribution.
   d. randomly spaced, evenly spaced, and unevenly spaced.

_____ 2. In the exponential model of population growth, the growth rate
   a. remains constant.   c. rises.
   b. declines.   d. rises and then declines.

_____ 3. *K*-strategists tend to live in environments that are
   a. unstable.   c. unpredictable.
   b. rapidly changing.   d. stable and predictable.

_____ 4. The Hardy-Weinberg principle
   a. can predict genotype frequencies affected by evolutionary forces.
   b. can predict genetic drift.
   c. applies only to large populations with nonrandom mating.
   d. Both (a) and (b)

_____ 5. In large, randomly mating populations, the frequencies of alleles and genotypes remain constant from generation to generation unless
   a. evolutionary forces are absent.
   b. evolutionary forces act on the population.
   c. the populations are *K*-strategists.
   d. the populations are *r*-strategists.

_____ 6. Natural selection acts on which of the following?
   a. genotypes   c. both phenotypes and genotypes
   b. phenotypes   d. neither phenotypes nor genotypes

_____ 7. Human height is an example of a
   a. single-gene trait.   c. monogenic trait.
   b. double-gene trait.   d. polygenic trait.

_____ 8. The range of phenotypes shifts toward one extreme in
   a. stabilizing selection.   c. directional selection.
   b. disruptive selection.   d. polygenic selection.

_____ 9. In a logistical model, exponential growth is limited by
   a. a density-independent factor.   c. a density-dependent factor.
   b. an unknown factor.   d. an exponential factor.

Copyright © by Holt, Rinehart and Winston. All rights reserved.

Holt Biology   Populations

Name _____ Class _____ Date _____

Test Prep Pretest *continued*

_____ **10.** All of the individuals of a species that live together in one place at one time are
  a. *K*-strategists.
  b. a population.
  c. a density-dependent factor.
  d. *r*-strategists.

In the space provided, write the letter of the description that best matches the term or phrase.

_____ **11.** logistic

_____ **12.** mutation

_____ **13.** natural selection

_____ **14.** *r*-strategists

_____ **15.** stabilizing selection

a. short lifespans and many offspring

b. frequencies of the intermediate phenotypes increase

c. a model of population growth that assumes that birth rates and death rates vary with population size

d. rates in nature are very slow

e. does not operate on rare, recessive alleles that are not expressed

Complete each statement by writing the correct term or phrase in the space provided.

16. To predict how a population will grow, demographers construct a(n) _____ of a population, a hypothetical population with the key characteristics of the real population being studied.

17. Organisms that produce few offspring that mature slowly are called _____ .

18. A female robin who chooses a male based on how well he sings is demonstrating _____ _____ .

19. Migration to or from a population results in _____ _____ .

20. If the graph of the phenotypes of a trait in a population is a hill-shaped curve, the trait exhibits a(n) _____ _____ .

21. When a recessive allele is present at a frequency of 0.1, only 1 out of 100 individuals will be homozygous recessive and will display the phenotype associated with this allele. However, 18 out of 1,000 individuals will be _____ and will carry the allele unexpressed.

Holt Biology 22 Populations

Name _____ Class _____ Date _____

Test Prep Pretest continued

Read each question, and write your answer in the space provided

22. Why does natural selection slowly reduce the frequency of harmful recessive alleles?

_____
_____
_____
_____
_____
_____

Questions 23 and 24 refer to the equations below.

A. $\Delta N = rN$

B. $\Delta N = rN \dfrac{(K - N)}{K}$

23. What number is being calculated in equation A?

_____

24. What happens when N approaches K in equation B?

_____
_____

Question 25 refers to the figures below.

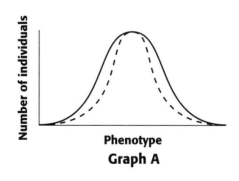

Phenotype
Graph A

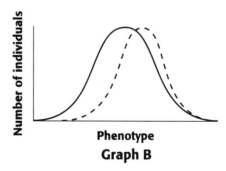

Phenotype
Graph B

25. What type of distribution does the solid-line curve in each of the graphs above represent?

_____

Name _____ Class _____ Date _____

Assessment
Quiz

Section: How Populations Grow

In the space provided, write the letter of the term or phrase that best completes each statement or best answers each question.

_____ 1. Populations tend to grow because
   a. the more of a species there are, the more likely they will survive.
   b. random events or natural disturbances are rare.
   c. individuals tend to have multiple offspring over their lifetime.
   d. there are always plentiful resources in every environment.

_____ 2. A school of fish in the ocean, when seen from a distance, displays which type of dispersion?
   a. clumped
   b. even
   c. random
   d. None of the above

_____ 3. Which of the following is a density-dependent factor that may limit population growth?
   a. climate change
   b. forest fire
   c. habitat destruction
   d. spread of disease

_____ 4. Which of the following equations can be used to predict the logistic growth rate of a population?
   a. $p^2 + 2pq + q^2 = 1$
   b. $\Delta N = rN\left[(K - N) \div K\right]$
   c. $\Delta N = rN$
   d. $r = \text{birthrate} - \text{death rate}$

_____ 5. Humans are *K*-strategists. Why does the human population continue to grow exponentially?
   a. Humans are capable of changing their environments to suit their needs.
   b. Unlike any other species, humans can occupy almost any habitat on Earth.
   c. Humans have not yet reached their carrying capacity.
   d. All of the above

Copyright © by Holt, Rinehart and Winston. All rights reserved.
Holt Biology   Populations

Name _____ Class _____ Date _____

Quiz continued

In the space provided, write the letter of the description that best matches the term or phrase.

_____ 6. population

_____ 7. dispersion

_____ 8. carrying capacity

_____ 9. exponential growth

_____ 10. logistic growth

a. population growth is limited by a density-dependent factor

b. all the individuals of a species that live together in one place at one time

c. the way the individuals of a population are arranged in space

d. the population size an environment can sustain over time

e. the rate of population growth stays the same, and population size increases steadily

Name _____ Class _____ Date _____

Assessment

Quiz

Section: How Populations Evolve

In the space provided, write the letter of the term or phrase that best completes each statement or best answers each question.

_____ 1. The Hardy-Weinberg principle states that
   a. dominant alleles will replace recessive alleles over time.
   b. dominant alleles are more common than recessive alleles.
   c. allele frequencies do not spontaneously change.
   d. evolutionary forces have no effect on allele frequencies.

_____ 2. Which of the following is NOT an evolutionary force?
   a. mutations in DNA
   b. Hardy-Weinberg principle
   c. genetic drift in a small population
   d. natural selection

_____ 3. Why are recessive alleles slow to be eliminated from a population?
   a. Recessive homozygotes are the only individuals that express the allele.
   b. Expressed recessive alleles are rare.
   c. Only expressed alleles can be acted on by natural selection.
   d. All of the above

_____ 4. Polygenic traits tend to exhibit a range of phenotypes clustered around an average value in a normal distribution because
   a. a trait can be influenced by many genes.
   b. a normal distribution curve takes many traits into account.
   c. a single gene controls many different traits.
   d. a distribution curve plots the traits of a single individual.

_____ 5. Examine the graph below. Which type of selection is apparent if the average value of the normal distribution is $x = 25$?

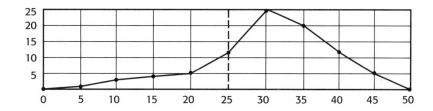

   a. stabilizing selection   **c.** directional selection
   b. natural selection   **d.** None of the above

Name _____ Class _____ Date _____

Quiz *continued*

In the space provided, write the letter of the description that best matches the term or phrase.

_____ 6. gene flow

_____ 7. mutation

_____ 8. nonrandom mating

_____ 9. genetic drift

_____ 10. natural selection

a. A bird prefers to breed with a male that is a strong nest builder.

b. Ten members leave a population and three others join.

c. A change in DNA results in the absence of an amino acid.

d. Over time, the allele for sickle cell anemia is decreasing in the United States because it provides no advantage.

e. A small group of ferrets, some with rare pink noses, is separated from the main population, and they begin to reproduce more pink-nosed ferrets.

Name _____ Class _____ Date _____

Assessment

Chapter Test

Populations

In the space provided, write the letter of the description that best matches the term or phrase.

_____ 1. genetic drift

_____ 2. *K*-strategists

_____ 3. small populations

_____ 4. even dispersal

_____ 5. recessive disorders

a. population increases slowly because of the production of few offspring

b. decreased allele frequency happens very slowly by natural selection

c. can result from a small group being separated from the main population

d. at risk of becoming extinct

e. located at regular intervals

In the space provided, write the letter of the term or phrase that best completes each statement or best answers each question.

_____ 6. Which of the following does NOT represent a population?
  a. all the robins in Austin, Texas
  b. all the grass frogs in a pond in Central Park, New York City
  c. all the birds in Chicago, Illinois
  d. all the black bears in Yosemite National Park

_____ 7. Actual proportions of homozygotes and heterozygotes can differ from Hardy-Weinberg predictions because of
  a. the occurrence of mutations.
  b. nonrandom mating among individuals.
  c. genetic drift within the population.
  d. All of the above

_____ 8. The movement of alleles into or out of a population is called
  a. mutation.
  b. gene flow.
  c. nonrandom mating.
  d. natural selection.

_____ 9. Cystic fibrosis is
  a. expressed in heterozygous individuals.
  b. caused by dominant alleles.
  c. a polygenic trait.
  d. expressed in homozygous individuals.

_____ 10. Natural selection shapes populations by acting on
  a. genes.
  b. recessive alleles.
  c. phenotypes.
  d. All of the above

Chapter Test continued

_____ 11. The type of selection that eliminates one extreme from a range of phenotypes is called
  a. directional selection.
  b. disruptive selection.
  c. polygenic selection.
  d. stabilizing selection.

_____ 12. Directional selection is characteristic of
  a. populations living in environments that do not change much.
  b. the evolution of single-gene traits.
  c. intermediate phenotypes.
  d. None of the above

Questions 13–15 refer to the figure below, which shows population growth over time.

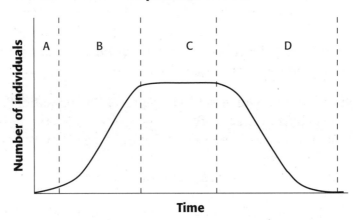

_____ 13. Which time period shows exponential growth of the population?
  a. A
  b. B
  c. C
  d. D

_____ 14. The rate of growth of a population is represented by r. During which time period will r equal zero?
  a. A
  b. B
  c. C
  d. D

_____ 15. The population size that can be sustained by an environment over time is
  a. a population curve.
  b. its carrying capacity.
  c. determined by allele frequency.
  d. exponential growth.

Chapter Test continued

In the space provided, write the letter of the description that best matches the term or phrase.

_____ 16. exponential growth

_____ 17. parental care

_____ 18. stabilizing selection

_____ 19. genetic uniformity

_____ 20. Hardy-Weinberg principle

a. results from genetic drift and can decrease a population's disease resistance

b. characteristic of *r*-strategists, followed by sudden decreases in population size

c. does not apply when evolutionary forces act on a population

d. range of phenotypes becomes narrower; more individuals in the middle range

e. characteristic of *K*-strategists, including a long period of development of offspring

Name _____ Class _____ Date _____

Assessment
Chapter Test

Populations

In the space provided, write the letter of the term or phrase that best completes each statement or best answers each question.

_____ 1. Weather and climate are environmental conditions that affect populations, and are known as
   a. density-dependent factors.
   b. density-independent factors.
   c. logistical factors.
   d. dispersion factors.

_____ 2. In a population study, r is found to have a negative value, which means that the population is
   a. increasing.
   b. staying the same.
   c. decreasing.
   d. extinct.

_____ 3. Populations of species that are r-strategists are characterized by
   a. long life spans.
   b. few offspring.
   c. large size.
   d. exponential growth.

_____ 4. Barnacles crowded together on a rock exhibit a type of dispersion called
   a. random dispersion.
   b. even dispersion.
   c. clumped dispersion.
   d. None of the above

_____ 5. Carrying capacity is
   a. the population size an environment can sustain over time.
   b. the maximum number of offspring a female can support at one time.
   c. the amount of allele frequency change that is possible because of natural selection.
   d. the number of polygenic traits in a population at one time.

_____ 6. The statistical study of populations is called
   a. growth curves.
   b. demography.
   c. geography.
   d. K-strategy.

_____ 7. The logistic model of population growth is usually associated with
   a. r-strategists.
   b. plants and insects.
   c. K-strategists.
   d. cockroaches and mosquitoes.

Copyright © by Holt, Rinehart and Winston. All rights reserved.
Holt Biology   Populations

Chapter Test *continued*

_____ 8. The Hardy-Weinberg principle applies to all populations as long as evolutionary forces are not working, and
   a. the population is large enough that its members are not likely to mate with one another.
   b. the ratio of genotypes and phenotypes differ significantly from each other.
   c. genotypes can be predicted using an equation.
   d. frequencies of alleles are not changing.

_____ 9. Not all mutations result in
   a. genetic changes.
   b. phenotypic changes.
   c. genotypic changes.
   d. codon changes.

_____ 10. Cheetahs are in danger of extinction because of the effects of
   a. mutations.
   b. genetic drift.
   c. gene flow.
   d. natural selection.

_____ 11. Allele frequency is least affected by
   a. genetic drift.
   b. gene flow.
   c. mutations.
   d. nonrandom mating.

_____ 12. In the United States, the allele for sickle cell anemia is
   a. quickly declining in frequency.
   b. rapidly increasing in frequency.
   c. neither increasing nor decreasing in frequency.
   d. slowly decreasing in frequency.

_____ 13. A demographer studying the adult height in males finds that more men are of average height now than 100 years ago, and there are fewer men who are very short or very tall. Which of the following may explain this trend?
   a. directional selection
   b. genetic drift
   c. stabilizing selection
   d. gene flow

_____ 14. In the Hardy-Weinberg equation, $p^2 + 2pq + q^2 = 1$, what does the term $2pq$ represent?
   a. frequency of homozygous dominant individuals
   b. frequency of heterozygous individuals
   c. frequency of homozygous recessive individuals
   d. None of the above

_____ 15. When directional selection eliminates one extreme from a range of phenotypes, the alleles promoting the extreme trait
   a. increase in the population.
   b. do not change from generation to generation.
   c. become less common in the population.
   d. None of the above

Name _____ Class _____ Date _____

Chapter Test continued

In the space provided, write the letter of the description that best matches the term or phrase.

_____ 16. gene flow

_____ 17. inbreeding

_____ 18. genetic drift

_____ 19. mutation

_____ 20. natural selection

_____ 21. directional selection

_____ 22. stabilizing selection

a. can result from a fire or landslide that leaves only a few survivors of a population

b. increases the proportion of similar individuals in a population

c. results from immigration and emigration

d. eliminates extremes from a range of phenotypes

e. does not significantly change allele frequencies except over very long periods of time

f. increases the proportion of homozygotes in a population

g. one of the most powerful agents of genetic change

Read each question, and write your answer in the space provided.

23. If you were determining how the size of a population might change, what four features of the population would you examine?

_____
_____
_____
_____

24. What type of curve is produced when N in the equation $\Delta N = rN$ is plotted against time on a graph?

_____
_____

25. Explain how stabilizing selection decreases genetic diversity.

_____
_____
_____
_____
_____
_____

Name _____ Class _____ Date _____

Quick Lab

DATASHEET FOR IN-TEXT LAB

Demonstrating the Hardy-Weinberg Principle

You can model the allele frequencies in a population with this simple exercise.

MATERIALS
- equal numbers of cards marked A or a to represent the dominant and recessive alleles for a trait
- paper bag

Procedure

1. You will be using the data table provided to record your data.
2. Work in a group, which will represent a population. Count the individuals in your group, and obtain that number of both A and a cards.
3. Place the cards in a paper bag, and mix them. Have each individual draw two cards, which represent a genotype. Record the genotype and phenotype in your data table.

Data Table					
	Trial 1	Trial 2	Trial 3	Trial 4	Trial 5
Genotype					
Phenotype					

4. Randomly exchange one "allele" with another individual in your group. Record the resulting genotypes.
5. Repeat Step 4 four more times.

Analysis

1. **Determine** the genotype and phenotype ratios in your group for each trial. Do the ratios vary among the trials?

   _____
   _____
   _____

2. **Hypothesize** what could cause a change in the "genetic makeup" of your group? Test one of your hypotheses.

   _____
   _____
   _____

Copyright © by Holt, Rinehart and Winston. All rights reserved.

Name _____ Class _____ Date _____

Math Lab

DATASHEET FOR IN-TEXT LAB

Building a Normal Distribution Curve

Background
You can help your class build a normal distribution curve by measuring the length of your shoes and plotting the data.

MATERIALS
- paper
- pencil
- measuring tape
- graph paper

Procedure

1. You will be using the data table provided to record your data.
2. Measure and record the length of one of your shoes to the nearest centimeter. Record your measurement and your gender.

Data Table			
Shoe length (cm)	Gender	Shoe length (cm)	Gender

3. Formulate a hypothesis about whether female shoes as a group are longer, shorter, or the same as shoes from males.
4. Determine the number of shoes of each length represented in the class.
5. Using a sheet of graph paper, make a graph showing the distribution of shoe length in your class. Show the number of students on the y-axis and shoe length on the x-axis.
6. Make a second graph using data only from females.
7. Make a third graph using data only from males.

Copyright © by Holt, Rinehart and Winston. All rights reserved.

Holt Biology — Populations

Name _____ Class _____ Date _____

Building a Normal Distribution Curve *continued*

Analysis

1. **Describe** the shape of the curve that resulted from the graph you made in step 5.

   _____

   _____

2. **Distinguish** how the distribution curve for shoe length of females differs from the curve for the shoe length of males.

   _____

   _____

   _____

3. **Predict** how the distribution curve that you made in step 5 would change if the data for males was deleted.

   _____

   _____

   _____

Exploration Lab
Observing How Natural Selection Affects a Population

DATASHEET FOR IN-TEXT LAB

SKILLS
- Using scientific methods
- Collecting, graphing, and analyzing data

OBJECTIVES
- **Measure** and collect data for a trait in a population.
- **Graph** a frequency distribution curve of your data.
- **Analyze** your data by determining its mean, median, mode, and range.
- **Predict** how natural selection can affect the variation in a population.

MATERIALS
- metric ruler
- graph paper (optional)
- green beans or snow peas
- calculator
- balance

Before You Begin

Natural selection can occur when there is **variation** in a **population**. You can analyze the variation in certain traits of a population by determining the mean, median, mode, and range of the data collected on several individuals. The **mean** is the sum of all data values divided by the number of values. The **median** is the midpoint in a series of values. The **mode** is the most frequently occurring value. The **range** is the difference between the largest and smallest values. The variation in a characteristic can be visualized with a **frequency distribution curve**. Two kinds of natural selection—**stabilizing selection** and **directional selection**—can influence the frequency and distribution of traits in a population. This changes the shape of a frequency distribution curve. In this lab, you will investigate variation in fruits and seeds.

1. Write a definition for each boldface term in the paragraph above. Use a separate sheet of paper.

2. Based on the objectives for this lab, write a question you would like to explore about variation in green beans or snow peas.

_____

_____

Name _____ Class _____ Date _____

Observing How Natural Selection Affects a Population *continued*

Procedure

PART A: DESIGN AN EXPERIMENT

1. Work with the members of your lab group to explore one of the questions written for step 2 of **Before You Begin.** To explore the question, design an experiment that uses the materials listed for this lab.

   > **You Choose**
   >
   > As you design your experiment, decide the following:
   >
   > **a.** what question you will explore
   >
   > **b.** what hypothesis you will test
   >
   > **c.** which trait (length, color, weight, etc.) you will measure
   >
   > **d.** how you will measure the trait
   >
   > **e.** how many members of the population you will measure (keep in mind that the more data you gather, the more revealing your frequency distribution curve will be)
   >
   > **f.** what data you will record in your data table

2. Write a procedure for your experiment. Make a list of all the safety precautions you will take. Have your teacher approve your procedure and safety precautions before you begin the experiment.

   _____
   _____
   _____
   _____
   _____
   _____
   _____

3. Conduct your experiment.

PART B: CLEANUP AND DISPOSAL

4. Dispose of seeds in the designated waste containers. Do not put lab materials in the trash unless your teacher tells you to do so.

5. Clean up your work area and all lab equipment. Return lab equipment to its proper place. Wash your hands thoroughly before you leave the lab and after you finish all work.

Name _____ Class _____ Date _____

Observing How Natural Selection Affects a Population *continued*

Analyze and Conclude

1. **Summarizing Results** Make a frequency distribution curve of your data. Plot the trait you measured on the x-axis (horizontal axis) and the number of times that trait occurred in your population on the y-axis (vertical axis).

2. **Calculating** Determine the mean, median, mode, and range of the data for the trait you studied.

_____

_____

_____

3. **Analyzing Results** How does the mean differ from the mode in your population?

_____

_____

4. **Drawing Conclusions** What type of selection appears to have produced the type of variation observed in your experiment?

_____

5. **Evaluating Data** The graph below shows the distribution of wing length in a population of birds on an island. Notice that the mean and the mode are quite different. Is the mean always useful in describing traits in a population? Explain.

Distribution of Wing Length

Number of Individuals (y-axis) vs *Wing Length* (x-axis), with Mode and Mean labeled on a bimodal distribution.

_____

_____

_____

_____

_____

Copyright © by Holt, Rinehart and Winston. All rights reserved.

Holt Biology — Populations

Name _____ Class _____ Date _____

Observing How Natural Selection Affects a Population *continued*

6. **Forming Hypotheses** What type of selection (stabilizing or directional) would be indicated if the mean of a trait you measured shifted, over time, to the right of a frequency distribution graph?

   _____

7. **Further Inquiry** Write a new question about variation in populations that could be explored in another investigation.

   _____

   _____

   _____

   _____

Name _____ Class _____ Date _____

Quick Lab

Using Random Sampling

MODELING

Scientists are not usually able to count every organism in a population. One way to estimate the size of a population is to collect data by taking random samples. In this lab, you will look at how data obtained by random sampling compare with data obtained by an actual count.

OBJECTIVES

Estimate a population's size by using a random sampling method.

Count the actual population size.

Compare data from random sampling with the actual count.

Compute percentage error to determine the accuracy of random sampling.

MATERIALS

- containers, small (2)
- pencil or pen
- scissors
- sheet of paper

Procedure

1. Cut a sheet of paper into 20 slips.
2. Number 10 of the slips from 1 to 10. Put the slips in a container.
3. Label the remaining 10 slips from A to J, and put them in a second container.
4. Review the grid in **Figure 1.** It represents a meadow measuring 10 m on each side. Each grid segment is 1 m². Each black circle represents one sunflower plant.
5. Randomly remove one slip from each container. On a sheet of paper, write the number-letter combination you drew, and find the grid segment in **Figure 1** that matches that combination. Count the number of sunflower plants in that grid segment. Record this number next to the number-letter combination on your sheet. Return each slip to the appropriate container.
6. Repeat step 5 until you have data for 10 different grid segments. These 10 grid segments represent a sample. Gathering data from a randomly selected sample of a larger area or group is called random sampling.
7. Find the total number of sunflower plants for the 10-segment sample. Divide this number by 10 to determine the average number of sunflower plants per grid segment, using the random sampling method. Record this average in **Table 1.**

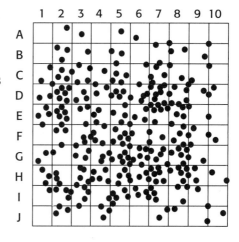

FIGURE 1 SUNFLOWER PLANTS IN A MEADOW

Copyright © by Holt, Rinehart and Winston. All rights reserved.

Holt Biology   45   Populations

Name _____ Class _____ Date _____

Using Random Sampling *continued*

8. Multiply the average number of plants per grid segment by 100 to find the estimated total number of plants in the meadow. Record this number in **Table 1**.

TABLE 1 NUMBER OF SUNFLOWER PLANTS IN A MEADOW

Method	Average number of plants per grid segment	Total number of plants in meadow
Random sampling		
Actual count		

9. Count the actual total number of sunflower plants in the meadow. Record this number in **Table 1**. It is the actual count. Divide the total number of sunflower plants by 100 to calculate the average number of plants per grid segment. Record this average in **Table 1**.

Analysis and Conclusions

1. **Explaining Events** Why was the paper-slip method used to select grid segments?

   _____

   _____

2. **Analyzing Results** What is the percentage error in your estimate of population size? To calculate percentage error, find the difference between the true value (the actual count) and the experimental value (the estimate obtained by random sampling). Divide this difference by the actual count and then multiply by 100.

$$\frac{\text{True value} - \text{Experimental value}}{\text{True value}} \times 100 = \% \text{ Error}$$

   _____

3. **Evaluating Methods** What does the percentage error indicate? How could you change the procedure in this lab to reduce your percentage error?

   _____

   _____

   _____

   _____

   _____

   _____

Name _____ Class _____ Date _____

Skills Practice Lab

Determining Growth Rate

CBL™ PROBEWARE

Populations do not grow indefinitely. Their sizes fluctuate, and are determined by *emigration* (leaving a population), *immigration* (joining a new population), *natality* (births), and *mortality* (deaths). Changes in the environment and the availability of resources impact these factors.

In this lab, you will investigate a population of yeast. The population will be closed, and can change only by natality and mortality. You will use a colorimeter to determine how rapidly the yeast population grows. A colorimeter determines the concentration of a solution by measuring the amount of absorbed light from a beam shining through a sample. You will learn how to construct a standard curve that will allow you to determine the concentration of yeast cells/mL from the measured absorbance.

OBJECTIVES

Use a colorimeter to measure the light absorbance of a growth medium containing a population of yeast cells.

Graph a growth curve for a population of yeast cells.

Determine the concentration of yeast cells in your growth medium from a standard curve of absorbance versus yeast concentration.

Compare the yeast population growth curve to a logistic growth curve.

MATERIALS

- apple juice (growth medium)
- CBL System
- colorimeter cuvette with lid
- colorimeter
- coverslip
- culture tube, 20 × 150 mm and cap
- graduated cylinder, 10 mL
- graph paper
- lab apron
- link cable
- microscope
- microscope slide
- paper towel
- pipets, 15 cm disposable (2)
- safety goggles
- stopper, test-tube
- test tube, 15 mm × 125 mm (2)
- test-tube rack
- TI graphing calculator
- water, distilled
- yeast solution

Copyright © by Holt, Rinehart and Winston. All rights reserved.

Holt Biology — Populations

Name _____ Class _____ Date _____

Determining Growth Rate *continued*

Procedure

PREPARING THE GROWTH MEDIUM

1. Put on safety goggles and a lab apron.

2. Label a clean, dry 20 × 150 mm culture tube with your name and lab period. Add 30 mL of apple juice to the tube.

3. Obtain the yeast solution from your teacher. Swish the solution gently to mix it evenly. Then extract some into a pipet. Add 3 mL of the yeast solution to the apple juice in the tube. Pipet the medium gently up and down a few times to mix the juice and yeast solution. Leave your tube in a warm location, such as on a shelf.

SETTING UP THE CBL SYSTEM

4. Connect the CBL unit and the calculator with the link cable. Press the link cable firmly into each device to assure a good connection. Connect the colorimeter to Channel 1 of the CBL unit.

5. Turn on the CBL unit and the calculator. Start the CHEMBIO program. Go to the MAIN MENU.

6. Select SET UP PROBES. Enter "1" as the number of probes. Select COLORIMETER from the SELECT PROBE menu. Enter "1" as the channel number. You are now ready to set up the growth medium and calibrate the CBL unit and colorimeter.

CALIBRATING THE COLORIMETER

7. To avoid bubble formation on the cuvette sides, slowly fill a colorimeter cuvette $\frac{3}{4}$ full of distilled water. Wipe the outside dry with a paper towel. If bubbles are present, gently tap the cuvette with your finger to loosen them.

8. Place the cuvette into the colorimeter with the ribs facing the front and back of the colorimeter chamber, as shown in **Figure 1**. A smooth side of the cuvette should be facing the white reference mark on the colorimeter.

FIGURE 1 PLACE THE CUVETTE IN THE COLORIMETER

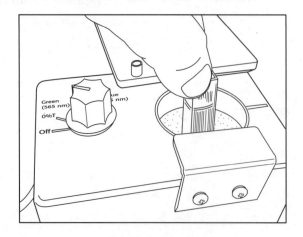

Determining Growth Rate *continued*

9. Close the colorimeter lid. Rotate the wavelength knob of the colorimeter to the "0 % T" position. When the display reading on the CBL unit is stable, press TRIGGER on the CBL unit.

10. On the graphing calculator, enter "0" as the reference. Turn the wavelength knob of the colorimeter to the Green LED position. When the CBL screen display is stable, press TRIGGER on the CBL unit. On the calculator, enter "100" as the reference. Leave the wavelength knob set to the Green LED setting to test samples.

11. On the calculator, press ENTER to return to the MAIN MENU. Select COLLECT DATA from the MAIN MENU. Select MONITOR INPUT from the DATA COLLECTION menu. On this setting, the CBL unit will monitor the light absorbance readings and display them on the graphing calculator.

MEASURING LIGHT ABSORBANCE

12. Agitate the culture tube gently to distribute the yeast organisms evenly. Remove the cap and withdraw about 2 mL of the growth medium with a sterile pipet. Fill a clean, dry cuvette with the growth medium from the pipet. Recap the culture tube immediately. Be sure the cuvette does not have any air bubbles in it. Wipe the outside dry with a paper towel. Rinse and dry the pipet.

13. Place the cuvette into the colorimeter. Close the lid and wait for the CBL display reading to stabilize. Read the absorbance value on the graphing calculator. If the absorbance reading is 1.0 or greater, follow the instructions in step 14 for diluting the growth medium sample. Record the absorbance reading and the dilution amount in **Table 1** for Day 1. If the growth medium was undiluted, record 1 for the dilution amount. If it was diluted by a factor of 10, record 0.1 for the dilution amount. If the first dilution was diluted, record 0.01 (1/100 dilution). Then, go to step 15.

14. Dilute your growth medium sample if the absorbance value on the colorimeter is greater than 1.0. Into a clean, dry test tube, place 1 mL of the growth medium from the cuvette. To the test tube containing the 1 mL of growth medium, add 9 mL of fresh juice. This produces a dilution of $\frac{1}{10}$ of the sample. Stopper the test tube, and shake gently to mix the contents. Remove 2 mL and place it into a clean cuvette. Wipe the cuvette and place it into the colorimeter. Follow the instructions in step 13 to measure the absorbance.

15. Turn off the CBL System by pressing "+" on the calculator, then QUIT. Remove the cuvette from the colorimeter, and save the solution to measure yeast concentration.

Name _____ Class _____ Date _____

Determining Growth Rate *continued*

MEASURING THE YEAST CONCENTRATION

16. Place a lid on the cuvette, and shake it gently to mix the medium evenly. Using a clean pipet, remove some medium from the cuvette. Place two drops on a microscope slide. Cover the drops with a clean coverslip. Put the slide on a microscope. Focus on low power, change to high power, and refocus as necessary. Adjust the light to view the yeast cells most easily.

17. Count the number of yeast cells visible in the high power (HP) field of view. A yeast cell with a bud counts as two. Record the number of yeast cells in **Table 1** under "Yeast counted." If time permits or if your teacher directs you to do so, move the slide to observe another section of it, or make a second slide. Then make another HP field count of yeast cells. Average the two counts, and record the average in **Table 1** under "Yeast counted."

18. Clean and dry the microscope slide and coverslip. Empty the cuvette into the sink, then rinse, and dry the cuvette. Clean up your work area and wash your hands. Store your labeled culture tube in a test-tube rack in a warm location.

19. Repeat steps 4–18 each day for up to 8 days or until you notice the absorbance values have peaked and are in decline.

TABLE 1 DAILY ABSORBANCE AND YEAST CONCENTRATION MEASUREMENTS

Day	Measured absorbance (%)	Dilution	Absorbance (%)	Yeast counted (no. of cells)	Actual yeast (no. of cells)	Cell concentration (no./mL)
1						
2						
3						
4						
5						
6						
7						
8						
9						

Determining Growth Rate continued

Analysis

1. **Organizing Data** Calculate the absorbance and actual yeast population in a high power field for each day. Record your values in **Table 1.** Compute absorbance by dividing the measured absorbance in **Table 1** by the dilution. If the dilution is 1.0, then the absorbance value will equal the measured absorbance value.

    Example: If the measured absorbance is 0.121 and the dilution is 0.1,

    $$\frac{\text{measured absorbance}}{\text{dilution}} = \frac{0.121}{0.1} = 1.21$$

    Compute the actual yeast population by dividing the diluted yeast population by the dilution.

    Example: If the diluted yeast population is 90 and the dilution is 0.1,

    $$\text{actual yeast population} = \frac{\text{diluted yeast population}}{\text{dilution}} = \frac{90}{0.1} = 900$$

2. **Organizing Data** Calculate cell concentration, using the data from **Table 1.** The cell concentration can be determined as follows:

    $$\text{no. yeast cells/mL} = \frac{\text{(actual number of cells in one HP field of view)}}{0.000011 \text{ mL}}$$

    Record the values of cell concentration in **Table 1.** (0.000011 mL is the volume in one field of view)

3. **Constructing Graphs** Use the data in **Table 1** to construct two graphs. In **Figure 2,** graph absorbance (*y*-axis) against the number of days (*x*-axis). In **Figure 3,** graph yeast cell concentration (*y*-axis) against absorbance (*x*-axis). Only use data prior to the decline of the growth rate. Draw a line of best fit in **Figure 3**.

FIGURE 2 ABSORBANCE VERSUS TIME

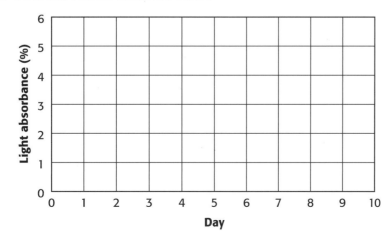

Determining Growth Rate *continued*

FIGURE 3 YEAST CONCENTRATION VERSUS ABSORBANCE

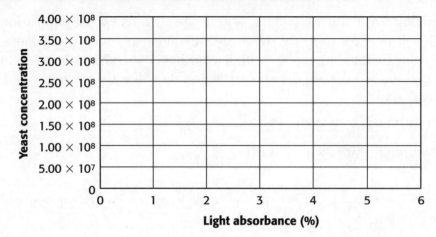

4. Analyzing Graphs Look at your graph of absorbance versus time. Describe the growth pattern of yeast during the experiment.

_____

_____

_____

5. Analyzing Graphs The concentration of growing yeast is important to biologists because they often need to use the cells while they are still growing exponentially. Look at your graph of absorbance versus time. During what range of absorbance are these cells growing exponentially?

_____

_____

_____

6. Identifying Relationships Biologists often use absorbance to determine the concentration of yeast and bacteria rather that doing actual cell counts each time. By using a standard curve such as the one you constructed in **Figure 3**, biologists can easily determine the concentration of microorganisms in growth media. What was the concentration of yeast in your sample when the absorbance was 1%? 2%?

_____

_____

_____

Determining Growth Rate *continued*

Conclusions

1. **Analyzing Graphs** Suppose you want to collect the most yeast you can for an experiment in which the yeast need to be growing exponentially. At which absorbance would you collect your yeast?

2. **Evaluating Methods** During the dilution process, why was juice used rather than water?

3. **Evaluating Results** If there are no limits to the growth of a population, the population will grow exponentially. But because there are limited resources in nature, a population's growth will slow, and fluctuate around a number called its *carrying capacity*. The carrying capacity is the maximum population size an environment can support for a long period of time. This kind of growth is called logistic growth, and is modeled by the *logistic growth curve*, shown in **Figure 4.** Does your graph in **Figure 2** look like the logistic model? In what ways is it similar or different?

FIGURE 4 THE LOGISTIC POPULATION GROWTH MODEL

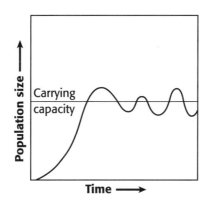

Name _____ Class _____ Date _____

Determining Growth Rate *continued*

4. **Drawing Conclusions** Why do you think your graph in Figure 2 does not level off at, or fluctuate around, a carrying capacity? (Hint: Suppose that the graph in **Figure 4** represents the population growth of moose on an island. Consider the differences between the growth of the moose population on the island and the yeast population in the media when accounting for any differences in the graphs.)

_____

_____

_____

_____

_____

5. **Drawing Conclusions** Describe the factors that contributed to the yeast population growth and decline.

_____

_____

_____

_____

_____

Extensions

1. Design an experiment to test and investigate the effects of other growth mediums on yeast growth. Other clear juices, sugar solutions, or Sabouraud dextrose broth may be good choices for other growth mediums.

2. Use the library, media center, or Internet to research the growth of the human population. Find out when and why the human population began to grow rapidly. Learn about the past and current human population sizes and draw a graph of human population growth. Compare it to the logistic model. Develop a visual presentation for your class. Include information about population growth in developed and developing countries, and human carrying capacity.

TEACHER RESOURCE PAGE

Name _____ Class _____ Date _____

Quick Lab

DATASHEET FOR IN-TEXT LAB

Demonstrating the Hardy-Weinberg Principle

You can model the allele frequencies in a population with this simple exercise.

MATERIALS
- equal numbers of cards marked A or a to represent the dominant and recessive alleles for a trait
- paper bag

Procedure
1. You will be using the data table provided to record your data.
2. Work in a group, which will represent a population. Count the individuals in your group, and obtain that number of both A and a cards.
3. Place the cards in a paper bag, and mix them. Have each individual draw two cards, which represent a genotype. Record the genotype and phenotype in your data table.

Data Table

	Trial 1	Trial 2	Trial 3	Trial 4	Trial 5
Genotype					
Phenotype					

4. Randomly exchange one "allele" with another individual in your group. Record the resulting genotypes.
5. Repeat Step 4 four more times.

Analysis
1. **Determine** the genotype and phenotype ratios in your group for each trial. Do the ratios vary among the trials?

   Unless all group members are heterozygous or all are homozygous (a rare situation, but more likely in small groups), the ratio of genotypes and phenotypes should remain constant for each trial.

2. **Hypothesize** what could cause a change in the "genetic makeup" of your group? Test one of your hypotheses.

   A new individual could join the group, or a member could leave (gene flow). Homozygous recessive individuals could be prevented from passing on their alleles (natural selection). Groups may also consider genetic drift and non-random mating.

TEACHER RESOURCE PAGE

Name _____ Class _____ Date _____

Math Lab — DATASHEET FOR IN-TEXT LAB
Building a Normal Distribution Curve

Background
You can help your class build a normal distribution curve by measuring the length of your shoes and plotting the data.

MATERIALS
- paper
- pencil
- measuring tape
- graph paper

Procedure
1. You will be using the data table provided to record your data.
2. Measure and record the length of one of your shoes to the nearest centimeter. Record your measurement and your gender.

Data Table			
Shoe length (cm)	Gender	Shoe length (cm)	Gender

3. Formulate a hypothesis about whether female shoes as a group are longer, shorter, or the same as shoes from males.
4. Determine the number of shoes of each length represented in the class.
5. Using a sheet of graph paper, make a graph showing the distribution of shoe length in your class. Show the number of students on the y-axis and shoe length on the x-axis.
6. Make a second graph using data only from females.
7. Make a third graph using data only from males.

Copyright © by Holt, Rinehart and Winston. All rights reserved.
Holt Biology

Building a Normal Distribution Curve continued

Analysis

1. **Describe** the shape of the curve that resulted from the graph you made in step 5.

   The graphs should resemble a hill-shaped curve.

2. **Distinguish** how the distribution curve for shoe length of females differs from the curve for the shoe length of males.

   Although the curves should appear similar, the peak of the graph of female shoe lengths should be to the left of (lower than) the peak of the graph of male lengths.

3. **Predict** how the distribution curve that you made in step 5 would change if the data for males was deleted.

   The curve would match the curve from step 5.

TEACHER RESOURCE PAGE

Name _____ Class _____ Date _____

Exploration Lab

DATASHEET FOR IN-TEXT LAB

Observing How Natural Selection Affects a Population

SKILLS
- Using scientific methods
- Collecting, graphing, and analyzing data

OBJECTIVES
- **Measure** and collect data for a trait in a population.
- **Graph** a frequency distribution curve of your data.
- **Analyze** your data by determining its mean, median, mode, and range.
- **Predict** how natural selection can affect the variation in a population.

MATERIALS
- metric ruler
- graph paper (optional)
- green beans or snow peas
- calculator
- balance

Before You Begin

Natural selection can occur when there is **variation** in a **population**. You can analyze the variation in certain traits of a population by determining the mean, median, mode, and range of the data collected on several individuals. The **mean** is the sum of all data values divided by the number of values. The **median** is the midpoint in a series of values. The **mode** is the most frequently occurring value. The **range** is the difference between the largest and smallest values. The variation in a characteristic can be visualized with a **frequency distribution curve**. Two kinds of natural selection—**stabilizing selection** and **directional selection**—can influence the frequency and distribution of traits in a population. This changes the shape of a frequency distribution curve. In this lab, you will investigate variation in fruits and seeds.

1. Write a definition for each boldface term in the paragraph above. Use a separate sheet of paper. **Answers appear in the TE for this lab.**

2. Based on the objectives for this lab, write a question you would like to explore about variation in green beans or snow peas.

   Answers will vary. For example: How much variation is there in the length of

   green beans?

Copyright © by Holt, Rinehart and Winston. All rights reserved.

Holt Biology — Populations

TEACHER RESOURCE PAGE

Name _____ Class _____ Date _____

Observing How Natural Selection Affects a Population *continued*

Procedure

PART A: DESIGN AN EXPERIMENT

1. Work with the members of your lab group to explore one of the questions written for step 2 of **Before You Begin.** To explore the question, design an experiment that uses the materials listed for this lab. **For Example: Measure the lengths or the weights of 20 green beans or snow peas.**

   > **You Choose**
   > As you design your experiment, decide the following:
   >
   > **a.** what question you will explore
   >
   > **b.** what hypothesis you will test
   >
   > **c.** which trait (length, color, weight, etc.) you will measure
   >
   > **d.** how you will measure the trait
   >
   > **e.** how many members of the population you will measure (keep in mind that the more data you gather, the more revealing your frequency distribution curve will be)
   >
   > **f.** what data you will record in your data table

2. Write a procedure for your experiment. Make a list of all the safety precautions you will take. Have your teacher approve your procedure and safety precautions before you begin the experiment.

   Organize the measurement values from the lowest to the highest value, and round to the nearest whole number. Beneath each value, write the number of green beans of that length. On the graph paper, make a graph showing the distribution curve.

3. Conduct your experiment. **Students should calculate the mean, median, and range for each type of measurement.**

PART B: CLEANUP AND DISPOSAL

4. Dispose of seeds in the designated waste containers. Do not put lab materials in the trash unless your teacher tells you to do so.

5. Clean up your work area and all lab equipment. Return lab equipment to its proper place. Wash your hands thoroughly before you leave the lab and after you finish all work.

Observing How Natural Selection Affects a Population *continued*

Analyze and Conclude

1. **Summarizing Results** Make a frequency distribution curve of your data. Plot the trait you measured on the x-axis (horizontal axis) and the number of times that trait occurred in your population on the y-axis (vertical axis).
   Most will aproximate a normal distribution.

2. **Calculating** Determine the mean, median, mode, and range of the data for the trait you studied.
   Answers will vary.

3. **Analyzing Results** How does the mean differ from the mode in your population?
   Most means will be similar to modes.

4. **Drawing Conclusions** What type of selection appears to have produced the type of variation observed in your experiment?
   Most will show stabilizing selection.

5. **Evaluating Data** The graph below shows the distribution of wing length in a population of birds on an island. Notice that the mean and the mode are quite different. Is the mean always useful in describing traits in a population? Explain.

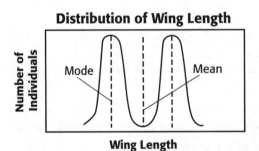

No. In the example given, the two modes in the population are very different from the mean. Thus, the mean is not always useful in describing traits in a population.

Name _____ Class _____ Date _____

Observing How Natural Selection Affects a Population continued

6. Forming Hypotheses What type of selection (stabilizing or directional) would be indicated if the mean of a trait you measured shifted, over time, to the right of a frequency distribution graph?

directional selection

7. Further Inquiry Write a new question about variation in populations that could be explored in another investigation.

For example: How does the size of a population of snow peas that has been grown in hot conditions compare to a population of snow peas grown in normal weather conditions?

TEACHER RESOURCE PAGE

Quick Lab

MODELING

Using Random Sampling

Teacher Notes

TIME REQUIRED 20 minutes

SKILLS ACQUIRED
Collecting data
Interpreting
Organizing and analyzing data

RATINGS Easy ←1——2——3——4→ Hard

Teacher Prep–1
Student Setup–2
Concept Level–3
Cleanup–1

THE SCIENTIFIC METHOD

Analyze the Results Analysis and Conclusions questions 1 and 2 require students to analyze their results.

TECHNIQUES TO DEMONSTRATE

To help students understand the meaning of percentage error, write the following formula on the board:

$$\frac{\text{True value} - \text{Experimental value}}{\text{True value}} \times 100 = \% \text{ Error}$$

Give students the following example: The standard boiling point of water at sea level is 100°C. A student measures the boiling point of water to be 99.6°C. What is the student's percentage error?

$$\frac{100 - 99.6}{100} \times 100 = 0.4\%$$

The student's percentage error was small and indicates a relatively accurate result. Ask students what a percentage error of 50% might indicate. (A high percentage error might indicate that the method the student used to collect the data was flawed or that the student was careless in some way.)

TIPS AND TRICKS

This lab works best in groups of two students.
    To save time, you might cut out and label the paper slips before the lab.
    For those circles that fall on the grid lines, have students include those that fall on the line at the top or the right of each square. Have students ignore those that fall on the line at the bottom or the left of each square.

Name _____ Class _____ Date _____

Quick Lab

Using Random Sampling

MODELING

Scientists are not usually able to count every organism in a population. One way to estimate the size of a population is to collect data by taking random samples. In this lab, you will look at how data obtained by random sampling compare with data obtained by an actual count.

OBJECTIVES

Estimate a population's size by using a random sampling method.

Count the actual population size.

Compare data from random sampling with the actual count.

Compute percentage error to determine the accuracy of random sampling.

MATERIALS

Procedure

1. Cut a sheet of paper into 20 slips.
2. Number 10 of the slips from 1 to 10. Put the slips in a container.
3. Label the remaining 10 slips from A to J, and put them in a second container.
4. Review the grid in **Figure 1**. It represents a meadow measuring 10 m on each side. Each grid segment is 1 m². Each black circle represents one sunflower plant.
5. Randomly remove one slip from each container. On a sheet of paper, write the number-letter combination you drew, and find the grid segment in **Figure 1** that matches that combination. Count the number of sunflower plants in that grid segment. Record this number next to the number-letter combination on your sheet. Return each slip to the appropriate container.
6. Repeat step 5 until you have data for 10 different grid segments. These 10 grid segments represent a sample. Gathering data from a randomly selected sample of a larger area or group is called random sampling.
7. Find the total number of sunflower plants for the 10-segment sample. Divide this number by 10 to determine the average number of sunflower plants per grid segment, using the random sampling method. Record this average in **Table 1**.

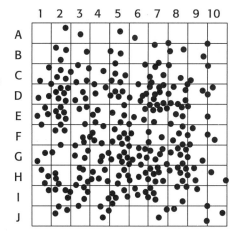

FIGURE 1 SUNFLOWER PLANTS IN A MEADOW

Copyright © by Holt, Rinehart and Winston. All rights reserved.

Holt Biology — Populations

Name _____ Class _____ Date _____

Using Random Sampling continued

8. Multiply the average number of plants per grid segment by 100 to find the estimated total number of plants in the meadow. Record this number in **Table 1**.

TABLE 1 NUMBER OF SUNFLOWER PLANTS IN A MEADOW

Method	Average number of plants per grid segment	Total number of plants in meadow
Random sampling	Data for the sampling method will depend on grid segments chosen.	
Actual count	26.1	261

9. Count the actual total number of sunflower plants in the meadow. Record this number in **Table 1**. It is the actual count. Divide the total number of sunflower plants by 100 to calculate the average number of plants per grid segment. Record this average in **Table 1**.

Analysis and Conclusions

1. **Explaining Events** Why was the paper-slip method used to select grid segments?

    Answers will vary but should suggest that the paper-slip method is a way to select the grid segments randomly.

2. **Analyzing Results** What is the percentage error in your estimate of population size? To calculate percentage error, find the difference between the true value (the actual count) and the experimental value (the estimate obtained by random sampling). Divide this difference by the actual count and then multiply by 100.

$$\frac{\text{True value} - \text{Experimental value}}{\text{True value}} \times 100 = \% \text{ Error}$$

3. **Evaluating Methods** What does the percentage error indicate? How could you change the procedure in this lab to reduce your percentage error?

    Answers will vary, but low percentage error indicates that random sampling can provide data that are reasonably close to actual counts. Students should recognize that making sure the sample is truly random and not localized in any way reduces percentage error. Also, increasing the number of random samples provides data that are closer to the actual counts, thereby reducing the percentage error.

TEACHER RESOURCE PAGE

Skills Practice Lab

CBL™ PROBEWARE

Determining Growth Rate

Teacher Notes

TIME REQUIRED One 45-minute period on the first day, 5–10 minutes per period on four to six observation days, 30 minutes on the final day

SKILLS ACQUIRED
Collecting data
Experimenting
Identifying and recognizing patterns
Interpreting
Measuring
Organizing and analyzing data

RATINGS
Easy ← 1   2   3   4 → Hard

Teacher Prep–4
Student Setup–3
Concept Level–2
Cleanup–2

THE SCIENTIFIC METHOD

Make Observations In Procedure steps 13, 14, 16, 17, and 19, students make observations of growth medium samples.

Analyze the Results Analysis questions 4–6 and Conclusions question 1 require students to analyze their results.

Draw Conclusions Conclusions questions 4 and 5 ask students to draw conclusions from their data.

MATERIALS

The yeast solution is made by mixing 1 g active dry yeast per 100 mL of distilled water. Freeze-dried yeast or baker's yeast may be used.

Other juices could be substituted for apple juice, but select one that is clear. To be sterile, apple juice must remain unopened until the time of use. An alternative to apple juice as a growth medium is Sabouraud dextrose broth available from WARD'S. See the *Master Materials List* for ordering instructions.

SAFETY CAUTIONS

• Discuss all safety symbols and caution statements with students.

• Caution students not to taste any of the yeast solution or juice growth medium.

DISPOSAL

Yeast solution can be rinsed down the drain under running water. Store-bought dry yeast can be mixed with water and rinsed down the drain. All other dry yeast should be destroyed by autoclaving or disinfecting with household bleach (5% sodium hypochlorite).

TEACHER RESOURCE PAGE

Determining Growth Rate continued

TECHNIQUES TO DEMONSTRATE

Demonstrate how to hold the colorimeter cuvette—by the top of the ribbed sides. This is the most secure way to hold the cuvette and minimizes chances for fingerprints on the smooth light-transmission surfaces.

TIPS AND TRICKS

Preparation

This lab works best in groups of two to four students.

The procedure in this lab is written for use with the original CBL system. If you are using CBL 2 or LabPro, the CHEMBIO program can still be used. Updated versions of this program can be downloaded from **www.vernier.com**. For additional information on how to integrate the CBL system into your laboratory, see the Program Introduction.

Some sensors may require the use of an adapter. Students will need to connect the adapter to the sensor before connecting it to the CBL.

To reduce observation time, calibrate the colorimeter and leave the CBL unit and calculator turned on during and between classes. Both the CBL unit and calculator will go into sleep mode. Press ON to turn them back on.

If you are using auto-ID colorimeters with CBL 2 or LabPro, the calibration instructions are slightly different from those in the lab procedure. Have students follow the on-screen instructions.

Yeast cell populations may be counted every other day and estimated from the final graph for days that were not counted. A correlation between the absorbance value and yeast cell population can be made.

Toward the end of the experiment, dead (lysed) yeast cell debris will be part of the mix, making the absorbance readings slightly higher than what would be expected with the cell count. For this reason, students are instructed in Analysis question 3 to disregard data once the growth rate is in decline, when making the graph in Figure 3.

Procedure

When students notice that the absorbance values have peaked and are in decline, typically in about 1 week, students can stop collecting data.

In step 7, students may use their growth medium (fruit juice) rather than distilled water for the colorimeter calibration. Try to use juice from the same container for consistency of color.

Students typically apply too much light when viewing yeast cells under a microscope, which makes the cells difficult to see. Yeast cells are most easily observed with low light levels. Stain the cells, if necessary, with Congo red.

Before removing a growth medium sample from the test tube, remind students to mix it well since the yeast cells tend to settle out.

As much as possible, readings should be taken at the same time each day.

The volume (in cm^3) of one field of view is calculated as follows: $\pi r^2 h$, where $\pi = 3.14$ and h is the height of the liquid between the slide and the coverslip, about 0.01 cm. Assume the volume to be 1.1×10^{-5} or 0.000011 mL.

TEACHER RESOURCE PAGE

Name _____ Class _____ Date _____

Skills Practice Lab

Determining Growth Rate

CBL™ PROBEWARE

Populations do not grow indefinitely. Their sizes fluctuate, and are determined by *emigration* (leaving a population), *immigration* (joining a new population), *natality* (births), and *mortality* (deaths). Changes in the environment and the availability of resources impact these factors.

In this lab, you will investigate a population of yeast. The population will be closed, and can change only by natality and mortality. You will use a colorimeter to determine how rapidly the yeast population grows. A colorimeter determines the concentration of a solution by measuring the amount of absorbed light from a beam shining through a sample. You will learn how to construct a standard curve that will allow you to determine the concentration of yeast cells/mL from the measured absorbance.

OBJECTIVES

Use a colorimeter to measure the light absorbance of a growth medium containing a population of yeast cells.

Graph a growth curve for a population of yeast cells.

Determine the concentration of yeast cells in your growth medium from a standard curve of absorbance versus yeast concentration.

Compare the yeast population growth curve to a logistic growth curve.

MATERIALS

- apple juice (growth medium)
- CBL System
- colorimeter cuvette with lid
- colorimeter
- coverslip
- culture tube, 20 × 150 mm and cap
- graduated cylinder, 10 mL
- graph paper
- lab apron
- link cable
- microscope
- microscope slide
- paper towel
- pipets, 15 cm disposable (2)
- safety goggles
- stopper, test-tube
- test tube, 15 mm × 125 mm (2)
- test-tube rack
- TI graphing calculator
- water, distilled
- yeast solution

Holt Biology — Populations

Name _____ Class _____ Date _____

Determining Growth Rate *continued*

Procedure

PREPARING THE GROWTH MEDIUM

1. Put on safety goggles and a lab apron.

2. Label a clean, dry 20 × 150 mm culture tube with your name and lab period. Add 30 mL of apple juice to the tube.

3. Obtain the yeast solution from your teacher. Swish the solution gently to mix it evenly. Then extract some into a pipet. Add 3 mL of the yeast solution to the apple juice in the tube. Pipet the medium gently up and down a few times to mix the juice and yeast solution. Leave your tube in a warm location, such as on a shelf.

SETTING UP THE CBL SYSTEM

4. Connect the CBL unit and the calculator with the link cable. Press the link cable firmly into each device to assure a good connection. Connect the colorimeter to Channel 1 of the CBL unit.

5. Turn on the CBL unit and the calculator. Start the CHEMBIO program. Go to the MAIN MENU.

6. Select SET UP PROBES. Enter "1" as the number of probes. Select COLORIMETER from the SELECT PROBE menu. Enter "1" as the channel number. You are now ready to set up the growth medium and calibrate the CBL unit and colorimeter.

CALIBRATING THE COLORIMETER

7. To avoid bubble formation on the cuvette sides, slowly fill a colorimeter cuvette $\frac{3}{4}$ full of distilled water. Wipe the outside dry with a paper towel. If bubbles are present, gently tap the cuvette with your finger to loosen them.

8. Place the cuvette into the colorimeter with the ribs facing the front and back of the colorimeter chamber, as shown in **Figure 1.** A smooth side of the cuvette should be facing the white reference mark on the colorimeter.

FIGURE 1 PLACE THE CUVETTE IN THE COLORIMETER

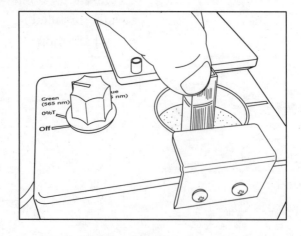

Determining Growth Rate *continued*

9. Close the colorimeter lid. Rotate the wavelength knob of the colorimeter to the "0 % T" position. When the display reading on the CBL unit is stable, press TRIGGER on the CBL unit.

10. On the graphing calculator, enter "0" as the reference. Turn the wavelength knob of the colorimeter to the Green LED position. When the CBL screen display is stable, press TRIGGER on the CBL unit. On the calculator, enter "100" as the reference. Leave the wavelength knob set to the Green LED setting to test samples.

11. On the calculator, press ENTER to return to the MAIN MENU. Select COLLECT DATA from the MAIN MENU. Select MONITOR INPUT from the DATA COLLECTION menu. On this setting, the CBL unit will monitor the light absorbance readings and display them on the graphing calculator.

MEASURING LIGHT ABSORBANCE

12. Agitate the culture tube gently to distribute the yeast organisms evenly. Remove the cap and withdraw about 2 mL of the growth medium with a sterile pipet. Fill a clean, dry cuvette with the growth medium from the pipet. Recap the culture tube immediately. Be sure the cuvette does not have any air bubbles in it. Wipe the outside dry with a paper towel. Rinse and dry the pipet.

13. Place the cuvette into the colorimeter. Close the lid and wait for the CBL display reading to stabilize. Read the absorbance value on the graphing calculator. If the absorbance reading is 1.0 or greater, follow the instructions in step 14 for diluting the growth medium sample. Record the absorbance reading and the dilution amount in **Table 1** for Day 1. If the growth medium was undiluted, record 1 for the dilution amount. If it was diluted by a factor of 10, record 0.1 for the dilution amount. If the first dilution was diluted, record 0.01 (1/100 dilution). Then, go to step 15.

14. Dilute your growth medium sample if the absorbance value on the colorimeter is greater than 1.0. Into a clean, dry test tube, place 1 mL of the growth medium from the cuvette. To the test tube containing the 1 mL of growth medium, add 9 mL of fresh juice. This produces a dilution of $\frac{1}{10}$ of the sample. Stopper the test tube, and shake gently to mix the contents. Remove 2 mL and place it into a clean cuvette. Wipe the cuvette and place it into the colorimeter. Follow the instructions in step 13 to measure the absorbance.

15. Turn off the CBL System by pressing "+" on the calculator, then QUIT. Remove the cuvette from the colorimeter, and save the solution to measure yeast concentration.

Determining Growth Rate continued

MEASURING THE YEAST CONCENTRATION

16. Place a lid on the cuvette, and shake it gently to mix the medium evenly. Using a clean pipet, remove some medium from the cuvette. Place two drops on a microscope slide. Cover the drops with a clean coverslip. Put the slide on a microscope. Focus on low power, change to high power, and refocus as necessary. Adjust the light to view the yeast cells most easily.

17. Count the number of yeast cells visible in the high power (HP) field of view. A yeast cell with a bud counts as two. Record the number of yeast cells in **Table 1** under "Yeast counted." If time permits or if your teacher directs you to do so, move the slide to observe another section of it, or make a second slide. Then make another HP field count of yeast cells. Average the two counts, and record the average in **Table 1** under "Yeast counted."

18. Clean and dry the microscope slide and coverslip. Empty the cuvette into the sink, then rinse, and dry the cuvette. Clean up your work area and wash your hands. Store your labeled culture tube in a test-tube rack in a warm location.

19. Repeat steps 4–18 each day for up to 8 days or until you notice the absorbance values have peaked and are in decline.

TABLE 1 DAILY ABSORBANCE AND YEAST CONCENTRATION MEASUREMENTS

Day	Measured absorbance (%)	Dilution	Absorbance (%)	Yeast counted (no. of cells)	Actual yeast (no. of cells)	Cell concentration (no./mL)
1	0.264	1	0.264	410	410	3.72×10^7
2	0.584	1	0.584	712	712	6.47×10^7
3	0.821	1	0.821	760	760	6.9×10^7
4	0.121	0.1	1.21	90	900	8.18×10^7
5	0.239	0.1	2.39	228	2280	2.07×10^8
6	0.516	0.1	5.16	400	4000	3.64×10^8
7	0.323	0.1	3.23	224	2240	2.04×10^8
8	0.310	0.1	3.10	144	1440	1.31×10^8
9	0.287	0.1	2.87	80	800	7.27×10^7

Data will vary. Sample data are entered above.

Determining Growth Rate continued

Analysis

1. **Organizing Data** Calculate the absorbance and actual yeast population in a high power field for each day. Record your values in **Table 1**. Compute absorbance by dividing the measured absorbance in **Table 1** by the dilution. If the dilution is 1.0, then the absorbance value will equal the measured absorbance value.

   Example: If the measured absorbance is 0.121 and the dilution is 0.1,

   $$\frac{\text{measured absorbance}}{\text{dilution}} = \frac{0.121}{0.1} = 1.21$$

   Compute the actual yeast population by dividing the diluted yeast population by the dilution.

   Example: If the diluted yeast population is 90 and the dilution is 0.1,

   $$\text{actual yeast population} = \frac{\text{diluted yeast population}}{\text{dilution}} = \frac{90}{0.1} = 900$$

2. **Organizing Data** Calculate cell concentration, using the data from **Table 1**. The cell concentration can be determined as follows:

   $$\text{no. yeast cells/mL} = \frac{(\text{actual number of cells in one HP field of view})}{0.000011 \text{ mL}}$$

   Record the values of cell concentration in **Table 1**. (0.000011 mL is the volume in one field of view)

3. **Constructing Graphs** Use the data in **Table 1** to construct two graphs. In **Figure 2**, graph absorbance (y-axis) against the number of days (x-axis). In **Figure 3**, graph yeast cell concentration (y-axis) against absorbance (x-axis). Only use data prior to the decline of the growth rate. Draw a line of best fit in **Figure 3**.

FIGURE 2 ABSORBANCE VERSUS TIME

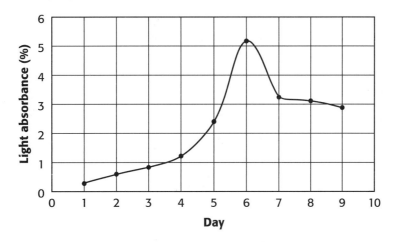

Name _____ Class _____ Date _____

Determining Growth Rate continued

FIGURE 3 YEAST CONCENTRATION VERSUS ABSORBANCE

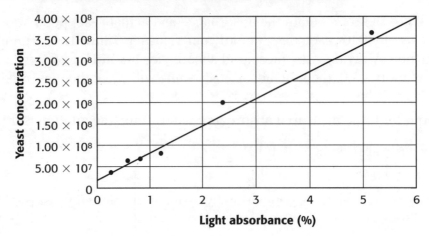

4. **Analyzing Graphs** Look at your graph of absorbance versus time. Describe the growth pattern of yeast during the experiment.

   Answers will vary. In general, students should observe a rapid increase in

   light absorbance and yeast concentration followed by a sharp decline.

5. **Analyzing Graphs** The concentration of growing yeast is important to biologists because they often need to use the cells while they are still growing exponentially. Look at your graph of absorbance versus time. During what range of absorbance are these cells growing exponentially?

   Answers will vary. Results should be consistent with students' graphs.

   According to the sample data, cells are growing exponentially in the light

   absorbance range from about 2–5 %.

6. **Identifying Relationships** Biologists often use absorbance to determine the concentration of yeast and bacteria rather that doing actual cell counts each time. By using a standard curve such as the one you constructed in **Figure 3**, biologists can easily determine the concentration of microorganisms in growth media. What was the concentration of yeast in your sample when the absorbance was 1%? 2%?

   According to the sample data, when the absorbance is 1%, the concentration

   is about 800 cells/mL. At an absorbance of 2%, the concentration is about

   1400 cells/mL.

Determining Growth Rate *continued*

Conclusions

1. **Analyzing Graphs** Suppose you want to collect the most yeast you can for an experiment in which the yeast need to be growing exponentially. At which absorbance would you collect your yeast?

    Answers will vary according to students' graphs. Sample data indicate that to

    collect the maximum amount of yeast, collect when the light absorbance

    reaches 5%.

2. **Evaluating Methods** During the dilution process, why was juice used rather than water?

    The juice matches the growth medium. The optical property (absorbance) of

    the juice will be the same as the growth medium.

3. **Evaluating Results** If there are no limits to the growth of a population, the population will grow exponentially. But because there are limited resources in nature, a population's growth will slow, and fluctuate around a number called its *carrying capacity*. The carrying capacity is the maximum population size an environment can support for a long period of time. This kind of growth is called logistic growth, and is modeled by the *logistic growth curve*, shown in **Figure 4**. Does your graph in **Figure 2** look like the logistic model? In what ways is it similar or different?

    The graph will probably not look like the logistic model. Students' graphs

    will most likely show an exponential rise and then will drop off steeply. The

    graphs should be similar to the logistic model in the exponential phase, but

    different from the logistic model in that the graphs continue to drop some-

    what instead of truly leveling off at a carrying capacity.

FIGURE 4 THE LOGISTIC POPULATION GROWTH MODEL

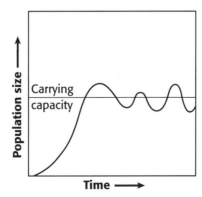

Name _____ Class _____ Date _____

Determining Growth Rate continued

4. Drawing Conclusions Why do you think your graph in Figure 2 does not level off at, or fluctuate around, a carrying capacity? (Hint: Suppose that the graph in **Figure 4** represents the population growth of moose on an island. Consider the differences between the growth of the moose population on the island and the yeast population in the media when accounting for any differences in the graphs.)

Answers will vary, but students should understand that, unlike the food

supply for the moose population, the food supply for the yeast population

was not replenished. This helps lead to a continued decline of the yeast

population rather than a steadying of the population around a carrying

capacity.

5. Drawing Conclusions Describe the factors that contributed to the yeast population growth and decline.

The growth of the yeast was due to a plentiful energy source and space to

grow. As the population became more crowded, food became scarce, waste

and cellular debris accumulated, and space was taken up. Because no further

food was provided, the population crashed.

Extensions

1. Design an experiment to test and investigate the effects of other growth mediums on yeast growth. Other clear juices, sugar solutions, or Sabouraud dextrose broth may be good choices for other growth mediums.
2. Use the library, media center, or Internet to research the growth of the human population. Find out when and why the human population began to grow rapidly. Learn about the past and current human population sizes and draw a graph of human population growth. Compare it to the logistic model. Develop a visual presentation for your class. Include information about population growth in developed and developing countries, and human carrying capacity.

Answer Key

Directed Reading

SECTION: HOW POPULATIONS GROW
1. population
2. demography
3. dispersion
4. size
5. Size is important because the number of individuals contained in a population can affect the population's ability to survive.
6. If there are too few individuals in an area, they may seldom encounter one another, making reproduction rare.
7. A population model is a hypothetical population that attempts to exhibit the key characteristics of a real population.
8. The growth rate is the difference between the death rate and the birth rate.
9. Density-dependent factors, such as food and water, refer to resources that are affected by the population density.
10. The logistic model is a population model in which exponential growth is limited by a density-dependent factor.
11. exponential growth curve
12. carrying capacity
13. b
14. a
15. c
16. d
17. a
18. b
19. c
20. a

SECTION: HOW POPULATIONS EVOLVE
1. d
2. e
3. b
4. c
5. a
6. b
7. c
8. c
9. a
10. a
11. Natural selection enables individuals that express favorable phenotypes to reproduce and pass those traits on to their offspring.
12. Hemophilia is caused by a recessive allele. Natural selection acts only on homozygous individuals, who express the trait.
13. polygenic
14. normal distribution
15. Directional selection is a type of selection in a population in which the frequency of a particular trait moves in one direction.
16. In stabilizing selection, the extremes at both ends of a range of phenotypes are eliminated, causing the frequencies of the intermediate phenotypes to increase.

Active Reading

SECTION: HOW POPULATIONS GROW
1. size, density, and dispersion
2. Very small populations are among those most likely to become extinct.
3. the number of individuals that live in a given area
4. When individuals of a population are spread widely apart, they have little opportunity for interactions. This hampers the reproductive capability of the population.
5. In a random distribution, the location of each individual is self-determined.
6. In an even distribution, individuals are located at regular intervals.
7. In a clumped distribution, individuals are bunched together in clusters.
8. a

SECTION: HOW POPULATIONS EVOLVE
1. Dominant alleles do not automatically replace recessive alleles.
2. the action of evolutionary forces
3. mutation, gene flow, nonrandom mating, genetic drift, and natural selection
4. They can cause the ratios of genotypes to differ significantly.
5. Mating with relatives because of small population size causes changes in the frequencies of alleles in that population.
6. d

Vocabulary Review

1. population
2. population size
3. population density
4. dispersion
5. population model
6. exponential growth curve
7. carrying capacity
8. density, dependent, factors
9. logistic model
10. density, dependent, factors
11. r-strategists
12. K-strategists
13. Hardy, Weinberg, principle
14. gene flow
15. nonrandom mating
16. genetic drift
17. polygenic trait
18. normal distribution
19. directional selection
20. stabilizing selection

Science Skills

INTERPRETING GRAPHS

1. The graph shows exponential growth. The population growth rate is constant, so the population size increases steadily.
2. This growth pattern is typical of r-strategists, such as bacteria and many plants and insects. When environmental conditions are favorable, these species reproduce rapidly. These species typically have short life spans, reproduce at a young age, and have many offspring each time they reproduce. The offspring are able to mature quickly with little or no parental care.
3. The Hardy-Weinberg principle applies to Population A. In a large population, many individuals carry this particular allele. The frequency of this allele may vary slightly but will remain relatively stable over generations.
4. The allele disappears completely. In such a small population, the allele was probably found in only a few individuals. Thus, the loss of even a few individuals through a chance mishap has a major effect on the frequency of the allele.

5. The Hardy-Weinberg principle applies to larger populations in which individuals mate randomly and larger populations in which evolutionary forces are not acting.
6. mutation, gene flow, nonrandom mating, genetic drift, and natural selection

Concept Mapping

1. population models
2. population size
3. population density
4. exponential growth curve
5. growth rate
6. density-independent factors
7. r-strategists
8. K-strategists
9. carrying capacity

Critical Thinking

1. d	13. f
2. b	14. h
3. a	15. j, h
4. f	16. a, l
5. e	17. k, c
6. c	18. f, d
7. b	19. b, i
8. g	20. g, e
9. d	21. b
10. e	22. b
11. c	23. d
12. a	

Quiz

SECTION: HOW POPULATIONS GROW

1. c	6. b
2. a	7. c
3. d	8. d
4. b	9. e
5. d	10. a

SECTION: HOW POPULATIONS EVOLVE

1. c	6. b
2. b	7. c
3. d	8. a
4. a	9. e
5. c	10. d

Test Prep Pretest

1. c	9. c
2. a	10. b
3. d	11. c
4. a	12. d
5. b	13. e
6. b	14. a
7. d	15. b
8. c	

16. model
17. *K*-strategists
18. nonrandom mating
19. gene flow
20. normal distribution
21. heterozygotes or *Aa*
22. Natural selection acts on phenotype, not genotype. Most individuals carrying the recessive allele are heterozygous and do not express it. Natural selection works only on the few individuals who are homozygous recessive and do express the allele. Thus, the frequency of such alleles is reduced very slowly.
23. the number of individuals that will be added to a population as it grows
24. The population ceases to grow because the birth rate equals the death rate.
25. a normal distribution

Chapter Test (General)

1. c	11. a
2. a	12. b
3. d	13. b
4. e	14. c
5. b	15. b
6. c	16. b
7. d	17. e
8. b	18. d
9. d	19. a
10. c	20. c

Chapter Test (Advanced)

1. b	12. d
2. c	13. c
3. d	14. b
4. c	15. c
5. a	16. c
6. b	17. f
7. c	18. a
8. a	19. e
9. b	20. g
10. b	21. d
11. c	22. b

23. population size, population density, growth rate, and dispersion
24. an exponential growth curve
25. Stabilizing selection results in fewer individuals in a population that have alleles promoting extreme types. Individuals become more similar, and genetic diversity decreases.

TEACHER RESOURCE PAGE

Lesson Plan

Section: How Populations Grow

Pacing

Regular Schedule: with lab(s): 5 days **without lab(s):** 3 days

Block Schedule: with lab(s): 2 1/2 days **without lab(s):** 1 1/2 days

Objectives

1. Distinguish among the three patterns of dispersion in a population.
2. Contrast exponential growth and logistic growth.
3. Differentiate *r*-strategists from *K*-strategists.

National Science Education Standards Covered

SCIENCE AS INQUIRY

SI1: Abilities necessary to do scientific inquiry

SI2: Understandings about scientific inquiry

LIFE SCIENCE: BIOLOGICAL EVOLUTION

LSEvol1: Species evolve over time.

LSEvol3: Natural selection and its evolutionary consequences provide a scientific explanation for the fossil record of ancient life forms as well as for the striking molecular similarities observed among the diverse species of living organisms.

LIFE SCIENCE: INTERDEPENDENCE OF ORGANISMS

LSInter4: Living organisms have the capacity to produce populations of infinite size, but environments and resources are finite.

LSInter5: Human beings live within the world's ecosystems.

LIFE SCIENCE: MATTER, ENERGY, AND ORGANIZATION IN LIVING SYSTEMS

LSMat5: The distribution and abundance of organisms and populations in ecosystems are limited by the availability of matter and energy and the ability of the ecosystem to recycle materials.

SCIENCE IN PERSONAL AND SOCIAL PERSPECTIVES

SPSP 2: Population growth

SPSP 3: Natural resources

TEACHER RESOURCE PAGE

Lesson Plan *continued*

KEY
SE = Student Edition TE = Teacher Edition
CRF = Chapter Resource File

Block 1

CHAPTER OPENER *(45 minutes)*

- **Quick Review,** SE. Students answer questions covered in previous sections of the textbook as preparation for the chapter content. (**GENERAL**)

- **Reading Activity,** SE. Before starting to read the chapter, students identify statements written on the board as true or false. Students find out if their predictions are correct after reading the chapter. (**GENERAL**)

- **Using the Figure,** TE. Students answer questions about the chapter opener photograph. (**GENERAL**)

- **Opening Activity**, Population Density, TE. Students calculate the population density of their local town or city. Ask students what information they need to get started, supply the information, and compare their results to the population density of other nearby towns as well as that of national and international cities. (**GENERAL**)

Block 2

FOCUS *(5 minutes)*

- **Bellringer Transparency.** Use this transparency as students enter the classroom and find their seats. (**GENERAL**)

MOTIVATE *(10 minutes)*

- **Activity,** Sampling, TE. Students outline a method of estimating the number of blades of grass in a local athletic field. (**GENERAL**)

TEACH *(30 minutes)*

- **Teaching Transparency, Section Outline.** Use this transparency to give students a framework for the information in this section. (**GENERAL**)

- **Teaching Transparency, Exponential Growth Curve.** Use this transparency to discuss exponential growth. Point out that exponential growth begins very slowly and accelerates as the number of reproducing individuals increases. (**GENERAL**)

- **Teaching Transparency, Logistic Growth.** Use this transparency to discuss logistic growth. Point out that logistic growth slows as resources become scarce and wastes accumulate. (**GENERAL**)

- **Teaching Tip,** What Values Can r Take?, TE. Explain to students that the value for r (growth rate) may be negative, positive, or 0. Then ask students which of these values the United States has. (**GENERAL**)

Copyright © by Holt, Rinehart and Winston. All rights reserved.

Holt Biology — Populations

TEACHER RESOURCE PAGE

Lesson Plan *continued*

HOMEWORK

- **Real Life**, SE. Students research the United States census. (**GENERAL**)
- **Active Reading Worksheet, How Populations Grow, CRF.** Students read a passage related to the section topic and answer questions. (**GENERAL**)
- **Directed Reading Worksheet, How Populations Grow, CRF.** Students complete the exercises in this worksheet to help them understand the material as they read the section. (**BASIC**)

Block 3

TEACH *(35 minutes)*

- **Teaching Tip,** Zero Population Growth, TE. Use the questions in this TE item to discucss zero population growth. (**GENERAL**)
- **Skill Builder,** Graphing, TE. Students graph historical population data for your area or town. (**GENERAL**)
- **Quick Lab, Using Random Sampling, CRF.** Students look at how data obtained by random sampling compare with data obtained by an actual count. (**BASIC**)

CLOSE *(10 minutes)*

- **Alternative Assessment**, TE. Students choose a familiar organism and indicate whether it is an example of an *r*-strategist species or a *K*-strategist species. (**GENERAL**)
- **Quiz**, TE. Students answer questions that review the section material. (**GENERAL**)

HOMEWORK

- **Reteaching,** TE. Students return to their lists of things they want to know about populations and make a list of what they have learned. (**BASIC**)
- **Section Review**, SE. Assign questions 1–5 for review, homework, or quiz. (**GENERAL**)
- **Quiz, CRF.** This quiz consists of ten multiple choice and matching questions that review the section's main concepts. (**BASIC**) **Also in Spanish.**

Optional Blocks

LAB *(45 minutes, plus 5–10 minutes per day for four to six observation days and 30 minutes on the final day)*

- **Skills Practice Lab, Determining Growth Rate, CRF.** Students examine and compare preserved specimens from the phylum Echinodermata (sea star, sea cucumber, sea dollar, and sea urchin) and the subphylums Urochordata (tunicate) and Cephalochordata (lancelet). (**GENERAL**)

Copyright © by Holt, Rinehart and Winston. All rights reserved.

TEACHER RESOURCE PAGE

Lesson Plan *continued*

Other Resource Options

- **Internet Connect.** Students can research Internet sources about Population Characteristics with SciLinks Code HX4143.
- **Internet Connect.** Students can research Internet sources about Population Growth Factors with SciLinks Code HX4145.
- **Internet Connect.** Students can research Internet sources about Population Pyramids with SciLinks Code HX4146.
- **go.hrw.com.** For worksheets, videos, and other teaching aids related to this chapter, visit the HRW Web site and type in the keyword HX4 EIC.
- **CNN Science in the News, Video Segment 12 Britain's Living Plan.** This video segment is accompanied by a **Critical Thinking Worksheet**.
- **Biology Interactive Tutor CD-ROM,** Unit 6 Ecosystem Dynamics. Students watch animations and other visuals as the tutor explains ecosystem dynamics. Students assess their learning with interactive activities.
- **CNN Student News.** Find the latest news, lesson plans, and activities related to important scientific events at **cnnstudentnews.com**.

TEACHER RESOURCE PAGE
Lesson Plan

Section: How Populations Evolve

Pacing
Regular Schedule: with lab(s): 4 days   without lab(s): 2 days
Block Schedule: with lab(s): 2 days   without lab(s): 1 day

Objectives
1. Summarize the Hardy-Weinberg principle.
2. Describe the five forces that cause a genetic change in a population.
3. Identify why selection against unfavorable recessive alleles is slow.
4. Contrast directional and stabilizing selection.

National Science Education Standards Covered

SCIENCE AS INQUIRY

SI1: Abilities necessary to do scientific inquiry

SI2: Understandings about scientific inquiry

LIFE SCIENCE: BIOLOGICAL EVOLUTION

LSEvol3: Natural selection and its evolutionary consequences provide a scientific explanation for the fossil record of ancient life forms as well as for the striking molecular similarities observed among the diverse species of living organisms.

LIFE SCIENCE: INTERDEPENDENCE OF ORGANISMS

LSInter4: Living organisms have the capacity to produce populations of infinite size, but environments and resources are finite.

LIFE SCIENCE: MATTER, ENERGY, AND ORGANIZATION IN LIVING SYSTEMS

LSMat5: The distribution and abundance of organisms and populations in ecosystems are limited by the availability of matter and energy and the ability of the ecosystem to recycle materials.

SCIENCE IN PERSONAL AND SOCIAL PERSPECTIVES

SPSP 2: Population growth

SPSP 3: Natural resources.

TEACHER RESOURCE PAGE

Lesson Plan *continued*

> **KEY**
> SE = Student Edition   TE = Teacher Edition
> CRF = Chapter Resource File

Block 4

FOCUS *(5 minutes)*

- **Bellringer Transparency.** Use this transparency as students enter the classroom and find their seats. **(GENERAL)**

MOTIVATE *(10 minutes)*

- **Identifying Preconceptions**, TE. Students discuss the frequency of polydactyly (having extra fingers) in humans.

TEACH *(30 minutes)*

- **Teaching Transparency, Section Outline.** Use this transparency to give students a framework for the information in this section. **(GENERAL)**

- **Exploring Further**, Using the Hardy-Weinberg Equation, SE. Students read the short article and then discuss the Hardy-Weinberg equation. **(GENERAL)**

- **Teaching Transparency, Two Kinds of Selection.** Use this transparency to dicuss directional and stabilizing selection. Help students relate the graphs to the concepts. **(GENERAL)**

- **Group Activity**, Genetic Drift: The Founder Effect, TE. Use this scenario to have students calculate the frequency of an allele in a population. **(GENERAL)**

HOMEWORK

- **Directed Reading Worksheet, How Populations Evolve, CRF.** Students complete the exercises in this worksheet to help them understand the material as they read the section. **(BASIC)**

- **Active Reading Worksheet, How Populations Evolve, CRF.** Students read a passage related to the section topic and answer questions. **(GENERAL)**

- **Problem Solving Worksheet, Genetics and Probability, CRF.** Students practice determining the possible offspring of genetic crosses and calculating the probability that each type of offspring will occur. **(GENERAL)**

Block 5

TEACH *(30 minutes)*

- **Quick Lab,** Demonstrating the Hardy-Weinberg Principle, SE. Students model allele frequency in a population using 3×5 in. cards. **(GENERAL)**

- **Datasheets for In-Text Labs,** Demonstrating the Hardy-Weinberg Principle, CRF.

Copyright © by Holt, Rinehart and Winston. All rights reserved.

TEACHER RESOURCE PAGE

Lesson Plan *continued*

- **Teaching Tip**, How Populations Evolve, TE. Students create a graphic organizer that represents the five forces that cause a population to evolve: mutation, gene flow, genetic drift, nonrandom mating, and natural selection. A sample graphic organizer is provided in the TE. (**GENERAL**)

CLOSE *(15 minutes)*

- **Reteaching**, TE. Tell students that a certain fictional trait (fuzzy wings) is caused by a recessive allele and prevents birds from flying swiftly. Ask them if they would expect rapid or slow selection against this allele and why. (**BASIC**)
- **Quiz**, TE. Students answer questions that review the section material. (**GENERAL**)

HOMEWORK

- **Math Lab,** Building a Normal Distribution Curve, SE. Students build a normal distribution curve using shoe-size measurements. (**GENERAL**)
- **Datasheets for In-Text Labs, Building a Normal Distribution Curve, CRF.**
- **Section Review,** SE. Assign questions 1–6 for review, homework, or quiz. (**GENERAL**)
- **Science Skills Worksheet, CRF.** Students interpret graphs showing changes in populations over time. (**GENERAL**)
- **Quiz, CRF.** This quiz consists of ten multiple choice and matching questions that review the section's main concepts. (**BASIC**) **Also in Spanish.**
- **Modified Worksheet, One-Stop Planner.** This worksheet has been specially modified to reach struggling students. (**BASIC**)
- **Critical Thinking Worksheet, CRF.** Students answer analogy-based questions that review the section's main concepts and vocabulary. (**ADVANCED**)

Optional Blocks

LAB *(90 minutes)*

- **Exploration Lab,** Observing How Natural Selection Affects a Population, SE. Students investigate variation in fruits and seeds. (**GENERAL**)
- **Datasheets for In-Text Labs, Observing How Natural Selection Affects a Population, CRF.**

Other Resource Options

- **Alternative Assessment,** TE. Use this scenario given in the TE to have students calculate allel frequencies. (**GENERAL**)
- **Supplemental Reading, Silent Spring, One-Stop Planner.** Students read the book and answer questions. (**ADVANCED**)
- **Internet Connect.** Students can research Internet sources about Hardy-Weinberg Equation with SciLinks Code HX4095.

TEACHER RESOURCE PAGE

Lesson Plan *continued*

- **go.hrw.com.** For worksheets, videos, and other teaching aids related to this chapter, visit the HRW Web site and type in the keyword HX4 EIC.
- **CNN Science in the News, Video Segment 12 Britain's Living Plan.** This video segment is accompanied by a **Critical Thinking Worksheet**.
- **Biology Interactive Tutor CD-ROM,** Unit 6 Ecosystem Dynamics. Students watch animations and other visuals as the tutor explains ecosystem dynamics. Students assess their learning with interactive activities.
- **CNN Student News.** Find the latest news, lesson plans, and activities related to important scientific events at **cnnstudentnews.com**.

TEACHER RESOURCE PAGE

Lesson Plan

End-of-Chapter Review and Assessment

Pacing

Regular Schedule: 2 days

Block Schedule: 1 day

KEY
SE = Student Edition   TE = Teacher Edition
CRF = Chapter Resource File

Block 6

REVIEW *(45 minutes)*

- **Study Zone,** SE. Use the Study Zone to review the Key Concepts and Key Terms of the chapter and prepare students for the Performance Zone questions. **(GENERAL)**
- **Performance Zone,** SE. Assign questions to review the material for this chapter. Use the assignment guide to customize review for sections covered. **(GENERAL)**
- **Teaching Transparency, Concept Mapping.** Use this transparency to review the concept map for this chapter. **(GENERAL)**

Block 7

ASSESSMENT *(45 minutes)*

- **Chapter Test, Populations, CRF.** This test contains 20 multiple choice and matching questions keyed to the chapter's objectives. **(GENERAL) Also in Spanish.**
- **Chapter Test, Populations, CRF.** This test contains 25 questions of various formats, each keyed to the chapter's objectives. **(ADVANCED)**
- **Modified Chapter Test, One-Stop Planner.** This test has been specially modified to reach struggling students. **(BASIC)**

Other Resource Options

- **Vocabulary Review Worksheet, CRF.** Use this worksheet to review the chapter vocabulary. **(GENERAL) Also in Spanish.**
- **Test Prep Pretest, CRF.** Use this pretest to review the main content of the chapter. Each question is keyed to a section objective. **(GENERAL) Also in Spanish.**
- **Test Item Listing for ExamView® Test Generator, CRF.** Use the Test Item Listing to identify questions to use in a customized homework, quiz, or test.
- **ExamView® Test Generator, One-Stop Planner.** Create a customized homework, quiz, or test using the HRW Test Generator program.

Copyright © by Holt, Rinehart and Winston. All rights reserved.

TEST ITEM LISTING
Populations

TRUE/FALSE

1. ____ The study of demographics helps predict changes in the size of a population.
   Answer: True   Difficulty: I   Section: 1   Objective: 1

2. ____ Very small populations are less likely to become extinct than larger populations.
   Answer: False   Difficulty: I   Section: 1   Objective: 1

3. ____ Wastes tend to accumulate in the environment as a population reaches the carrying capacity.
   Answer: True   Difficulty: I   Section: 1   Objective: 2

4. ____ Populations of *K*-strategists grow rapidly, while *r*-strategist populations grow slowly.
   Answer: False   Difficulty: I   Section: 1   Objective: 3

5. ____ The Hardy-Weinberg principle states that the proportions of recessive and dominant alleles in a population fluctuate randomly from generation to generation.
   Answer: False   Difficulty: I   Section: 2   Objective: 1

6. ____ Mutations are so common that they are the major cause of changes in allele frequencies within a population.
   Answer: False   Difficulty: I   Section: 2   Objective: 2

7. ____ Natural selection acts on phenotypes, not genotypes.
   Answer: True   Difficulty: I   Section: 2   Objective: 3

8. ____ Natural selection always eliminates any genetic disorders from a population, regardless of the frequency of the gene that is responsible for a disorder.
   Answer: False   Difficulty: I   Section: 2   Objective: 3

9. ____ Directional selection results in the range of phenotypes shifting toward one extreme.
   Answer: True   Difficulty: I   Section: 2   Objective: 4

10. ____ In stabilizing selection, the range of phenotypes becomes wider.
    Answer: False   Difficulty: I   Section: 2   Objective: 4

MULTIPLE CHOICE

11. Which of the following does *not* represent a population?
    a. all the robins in Austin, Texas
    b. all the grass frogs in the pond of Central Park, New York City
    c. all the birds in Chicago, Illinois
    d. all the earthworms in Yosemite National Park
    Answer: C   Difficulty: I   Section: 1   Objective: 1

12. Because individuals in a population usually tend to produce more than one offspring,
    a. populations tend to increase in size.
    b. populations remain stable in size.
    c. individuals tend to die quickly.
    d. the number of individuals declines rapidly.
    Answer: A   Difficulty: I   Section: 1   Objective: 1

TEST ITEM LISTING, continued

13. All of the following are problems arising from inbreeding *except*
    a. production of a genetically uniform population.
    b. increases in the diversity within a population.
    c. increased chance of homozygous recessive alleles occurring.
    d. reduction of a population's ability to adapt to environmental changes.
    Answer: B   Difficulty: I   Section: 1   Objective: 1

14. Demographic studies of populations must take into consideration
    a. population size.
    b. population density.
    c. population dispersion.
    d. All of the above
    Answer: D   Difficulty: I   Section: 1   Objective: 1

15. Regarding population dispersion patterns, which of the following is an *inappropriate* pairing?
    a. randomly spaced — chance
    b. evenly spaced — regular intervals
    c. clumped — clusters
    d. dispersive — randomly distributed
    Answer: D   Difficulty: I   Section: 1   Objective: 1

16. population density : number of individuals in a given area ::
    a. population : an area where organisms live
    b. logistic growth : how populations grow in nature
    c. logistic growth curve : exponential rate of growth
    d. population size : population density
    Answer: B   Difficulty: II   Section: 1   Objective: 1

17. As a population reaches its carrying capacity, there is an increase in competition for
    a. food.
    b. shelter.
    c. mates.
    d. All of the above
    Answer: D   Difficulty: I   Section: 1   Objective: 2

Population Growth Over Time

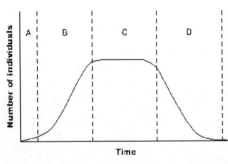

18. Refer to the illustration above. Which time period shows exponential growth of the population?
    a. period A
    b. period B
    c. period C
    d. period D
    Answer: B   Difficulty: II   Section: 1   Objective: 2

19. Refer to the illustration above. During which time period are the birth rate and death rate equal?
    a. period A
    b. period B
    c. period C
    d. period D
    Answer: C   Difficulty: II   Section: 1   Objective: 2

TEST ITEM LISTING, continued

20. Refer to the illustration above. The rate of growth of a population is represented by r. During which time period will $r = 0$?
    a. period A
    b. period B
    c. period C
    d. period D
    Answer: C   Difficulty: II   Section: 1   Objective: 2

21. Refer to the illustration above. The time period during which r (the rate of growth of a population) would have a negative value is
    a. period A.
    b. period B.
    c. period C.
    d. period D.
    Answer: D   Difficulty: II   Section: 1   Objective: 2

22. birth and death rates : constant on exponential growth curve ::
    a. birth rates : equal to death rates
    b. r-strategists : equal to K-strategists
    c. birth and death rates : not constant on logistic growth curve
    d. exponential models : same as logistic models
    Answer: C   Difficulty: II   Section: 1   Objective: 2

23. Environments that are unpredictable and rapidly changing tend to support populations of
    a. Q-strategists.
    b. K-strategists.
    c. N-strategists.
    d. r-strategists.
    Answer: D   Difficulty: I   Section: 1   Objective: 3

24. All of the following are true of r-strategists *except*
    a. early maturation and reproduction.
    b. little parental care.
    c. few offspring.
    d. small offspring.
    Answer: C   Difficulty: I   Section: 1   Objective: 3

25. Which of the following are r-strategists?
    a. redwoods
    b. dandelions
    c. whales
    d. humans
    Answer: B   Difficulty: I   Section: 1   Objective: 3

26. Which of the following are *inappropriately* paired?
    a. K-strategists — reproduce late in life
    b. K-strategists — minimal parental care
    c. r-strategists — reproduce early in life
    d. r-strategists — mature quickly
    Answer: B   Difficulty: I   Section: 1   Objective: 3

27. bacteria : r-strategists ::
    a. gorillas : K-strategists
    b. insects : K-strategists
    c. annual plants : K-strategists
    d. rhinoceroses : r-strategists
    Answer: A   Difficulty: II   Section: 1   Objective: 3

28. In 1908, Hardy and Weinberg independently demonstrated that
    a. r- and K-strategist populations are actually the same.
    b. recessive alleles replace dominant alleles in a population over long periods of time.
    c. dominant alleles do not replace recessive alleles in a population.
    d. recessive alleles are usually more common than dominant alleles.
    Answer: C   Difficulty: II   Section: 2   Objective: 1

29. Actual proportions of homozygotes and heterozygotes can differ from Hardy-Weinberg predictions because of
    a. the occurrence of mutations.
    b. nonrandom mating among individuals.
    c. genetic drift within the population.
    d. All of the above
    Answer: D   Difficulty: I   Section: 2   Objective: 1

TEST ITEM LISTING, continued

30. The movement of alleles into or out of a population due to migration is called
    a. mutation.
    b. gene flow.
    c. nonrandom mating.
    d. natural selection.

    Answer: B   Difficulty: I   Section: 2   Objective: 2

31. Inbreeding
    a. is a form of random mating.
    b. causes mutations to occur.
    c. increases the proportion of heterozygotes.
    d. increases the proportion of homozygotes.

    Answer: D   Difficulty: I   Section: 2   Objective: 2

32. nonrandom mating : increasing proportion of homozygotes ::
    a. migration of individuals : gene flow
    b. mutation : major change in allele frequencies
    c. Hardy-Weinberg equation : natural selection
    d. inbreeding : frequency of alleles

    Answer: A   Difficulty: II   Section: 2   Objective: 2

33. homozygous : heterozygous ::
    a. heterozygous : *Bb*
    b. probability : predicting chances
    c. dominant : recessive
    d. factor : gene

    Answer: C   Difficulty: II   Section: 2   Objective: 2

34. Natural selection acts
    a. only on heterozygous genotypes.
    b. only on recessive alleles.
    c. on phenotypes that are expressed.
    d. on all mutations.

    Answer: C   Difficulty: I   Section: 2   Objective: 3

35. Directional selection tends to eliminate
    a. both extremes in a range of phenotypes.
    b. one extreme in a range of phenotypes.
    c. intermediate phenotypes.
    d. None of the above; it causes new phenotypes to form.

    Answer: B   Difficulty: I   Section: 2   Objective: 4

36. The range of phenotypes shifts toward one extreme in
    a. stabilizing selection.
    b. disruptive selection.
    c. directional selection.
    d. polygenic selection.

    Answer: C   Difficulty: I   Section: 2   Objective: 4

COMPLETION

37. A(n) _____ consists of all the individuals of a particular species in a particular place.

    Answer: population   Difficulty: I   Section: 1   Objective: 1

38. The statistical study of all populations is called _____.

    Answer: demography   Difficulty: I   Section: 1   Objective: 1

39. Population density refers to how many _____ are present in a particular location.

    Answer: individuals   Difficulty: I   Section: 1   Objective: 1

40. The way in which members of a population are arranged in a given area is referred to as _____.

    Answer: dispersion   Difficulty: I   Section: 1   Objective: 1

TEST ITEM LISTING, continued

41. A population _____ is a hypothetical population that has key characteristics of the real population being studied.
    Answer: model   Difficulty: I   Section: 1   Objective: 2

42. The difference between the birth rate and death rate of a population is the _____ _____.
    Answer: growth rate   Difficulty: I   Section: 1   Objective: 2

43. The _____ _____ is the population size that can be sustained by an environment.
    Answer: carrying capacity   Difficulty: I   Section: 1   Objective: 2

44. Species that are _____ -strategists tend to have periods of exponential growth followed by sudden crashes in population size.
    Answer: r   Difficulty: I   Section: 1   Objective: 3

45. Small population sizes and slow population growth are typical of organisms that are _____ -strategists.
    Answer: K   Difficulty: I   Section: 1   Objective: 3

46. Stable and predictable environments are commonly inhabited by populations of organisms that are _____ -strategists.
    Answer: K   Difficulty: I   Section: 1   Objective: 3

47. Alternative versions of genes are called _____.
    Answer: alleles   Difficulty: I   Section: 2   Objective: 1

48. The movement of individuals from one population to another is called _____.
    Answer: migration   Difficulty: I   Section: 2   Objective: 2

49. Migration results in _____ _____.
    Answer: gene flow   Difficulty: I   Section: 2   Objective: 2

50. A characteristic influenced by several genes is called a(n) _____ trait.
    Answer: polygenic   Difficulty: I   Section: 2   Objective: 3

51. When the range of phenotypes becomes narrower, increasing the number of individuals with characteristics near the middle of the range, this is called _____ _____.
    Answer: stabilizing selection   Difficulty: I   Section: 2   Objective: 4

TEST ITEM LISTING, continued

ESSAY

The graph below depicts the growth of a population of fruit flies over time.

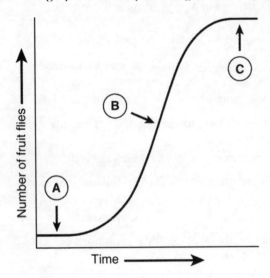

52. Would a density-dependent limiting factor have a greater impact on the population at point A, B, or C on the curve? Why?

    Answer: It would have the greatest impact on the population at point C. This is because point C on the curve indicates the greatest population density. Density-dependent limiting factors impact populations more as they increase in size.
    Difficulty: III   Section: 1   Objective: 1

53. Why does the population stop increasing after it reaches the point on the curve labeled C?

    Answer: The population has reached the carrying capacity of the ecosystem in which it lives. The ecosystem cannot support any more flies than this number over time.
    Difficulty: III   Section: 1   Objective: 2

54. Name one density-independent limiting factor that could affect this population of fruit flies. Would you expect this limiting factor to have a greater impact on the population at any particular point on the curve, and if so, which one?

    Answer: A number of abiotic factors would be suitable answers: temperature extremes (for example, freezing), floods, hurricanes, fires, and volcanic eruptions. Such limiting factors would not be expected to have a greater impact on the population at any particular point on the curve.
    Difficulty: III   Section: 1   Objective: 2

TEST ITEM LISTING, continued

55. Contrast exponential growth with logistical growth by completing the chart.

Criteria	Exponential Population Growth	Logistic Population Growth
Graph of Growth Rate		
Assumptions		
Birth and death Rates		

Answer:

Criteria	Exponential Population Growth	Logistic Population Growth
Graph of Growth Rate	*(J-shaped curve)*	*(S-shaped curve)*
Assumptions	unlimited resources	resources limit population growth; stabilizes at the carrying capacity
Birth and death Rates	constant—rates do not change	vary with population size

Difficulty: III   Section: 1   Objective: 2

56. Explain the difference between *r*-strategist and *K*-strategist populations.
    Answer:
    These two types of populations differ in their rates of maturation and reproduction, the amount of parental care of offspring, and the type of environments that they inhabit. Species that are *r*-strategists mature quickly, reproduce in large numbers, and invest little energy in care of their offspring. This adapts them for unpredictable, rapidly changing environments. *K*-strategists tend to be larger organisms that mature more gradually and exhibit parental care of a smaller number of offspring. *K*-strategists are found in more stable environments.
    Difficulty: III   Section: 1   Objective: 3

57. Distinguish between the two types of natural selection acting on polygenic traits.
    Answer:
    Directional selection tends to eliminate individuals in a population that are at one or the other extreme of the range of phenotypes. Stabilizing selection narrows the range of individuals to those in the middle of the range.
    Difficulty: II   Section: 2   Objective: 4